International Correspondence Schools

An Elementary Treatise on Electric Power and Lighting

Vol. 1, First Edition

International Correspondence Schools

An Elementary Treatise on Electric Power and Lighting
Vol. 1, First Edition

ISBN/EAN: 9783337249595

Printed in Europe, USA, Canada, Australia, Japan

Cover: Foto ©berggeist007 / pixelio.de

More available books at **www.hansebooks.com**

AN ELEMENTARY TREATISE

ON

ELECTRIC POWER AND LIGHTING

Prepared for Students of
THE INTERNATIONAL CORRESPONDENCE SCHOOLS
SCRANTON, PA.

Volume I

ARITHMETIC
MENSURATION
MECHANICS

WITH PRACTICAL QUESTIONS AND EXAMPLES

First Edition

SCRANTON
THE COLLIERY ENGINEER CO.
1897

Entered according to the Act of Congress, in the year 1897, by THE COLLIERY
ENGINEER COMPANY, in the office of the Librarian of Congress.
at Washington.

BURR PRINTING HOUSE,
FRANKFORT AND JACOB STREETS,
NEW YORK.

PREFACE.

The Instruction and Question Papers which are furnished to the students of The International Correspondence Schools become so badly worn and soiled that, when a student has completed his Course, he has worn out his Instruction Papers, and they are no longer suitable for reference or review. Since the Instruction Papers are very valuable, especially to those who have studied them, there has grown up a demand for Sets of the Instruction and Question Papers, indexed for convenient reference, and durably bound for preservation. Again, many of our students can spare but little time for study, and are, therefore, a long while passing through their Courses. These students also desire the Papers in bound volumes to use for reference. Other students begin Courses, but for various reasons are unable to complete them, and feel that, having paid for their Scholarships, they ought to have the text-books, even though they can not finish their Courses.

For these reasons, we have decided to publish all of the Instruction and Question Papers of our different Technical Courses in volumes bound in Half Leather, to make a small advance in our prices, and to furnish a set to each student as soon as his Scholarship is paid for, whether he has completed his studies or not.

The volumes for the present Course, the Electric Power and Lighting, are five in number:

Volume I contains the Instruction and Question Papers on Arithmetic, Mensuration, and Mechanics.

Volume II contains the Instruction and Question Papers on Dynamos and Motors, Electric Lighting, and Electric Railways.

PREFACE.

Volume III contains the Plates and the instructions for drawing them. It forms a very complete Course in Mechanical Drawing.

Volume IV contains the Tables, Rules, and Formulas in common use. The student who has finished his Course will find these Tables of great service. All the principal formulas, with the definitions of the letters used in them, and the principal rules, are conveniently arranged for reference, so that the student can save the labor and time of hunting them out in the Instruction Papers.

Volume V contains the answers to the questions in the Question Papers.

These Instruction Papers are written from a practical standpoint, and contain only such information as the student requires in order to obtain a good working knowledge of the subjects which form the groundwork of a course in Electric Power and Lighting. There is no padding, and the student gets in a clear and concise form the exact information which he desires.

To keep the student always interested in his work, we do not give him half a dozen ways of doing the same thing; neither do we enter into speculative discussions of different subjects, all of which tend to confuse the student and leave him in doubt as to which one (or any one) is correct; we give him the best or most suitable rule, formula, or method that we know of, without mentioning any of the others. In short, the Papers are written for practical men, and all the difficulties encountered by a student studying by himself, particularly those which are due to a definition or explanation being too technical, abstract, or not clear enough to be readily understood, have been carefully considered and overcome.

THE INTERNATIONAL CORRESPONDENCE SCHOOLS.

CONTENTS.

ARITHMETIC.   PAGE

    Definitions, - - - - - - - 1
    Notation and Numeration, - - - - - 1
    Addition, - - - - - - - - 4
    Subtraction, - - - - - - - - 9
    Multiplication, - - - - - - - 12
    Division, - - - - - - - - 17
    Cancelation, - - - - - - - - 21
    Fractions, - - - - - - - - 23
    Decimals, - - - - - - - - 38
    Percentage, - - - - - - - - 55
    Denominate Numbers, - - - - - - 61
    Involution, - - - - - - - - 76
    Evolution, - - - - - - - - 79
    Ratio, - - - - - - - - - 96
    Proportion, - - - - - - - - 100

MENSURATION AND USE OF LETTERS IN ALGEBRAIC FORMULAS.

    Formulas, - - - - - - - - 115
    Lines and Angles, - - - - - - - 119
    Quadrilaterals, - - - - - - - 123
    The Triangle, - - - - - - - 126
    Polygons, - - - - - - - - 130
    The Circle, - - - - - - - - 133
    The Prism and Cylinder, - - - - - 138
    The Pyramid and Cone, - - - - - 142
    The Frustum of a Pyramid or Cone, - - - 143
    The Sphere and Cylindrical Ring, - - - 145

MECHANICS.

    Matter and its Properties, - - - - - 149
    Motion and Velocity, - - - - - - 153
    Force, - - - - - - - - - 156

CONTENTS.

MECHANICS—*continued*. PAGE
Center of Gravity, - - - - - - 160
Simple Machines, - - - - - - 165
Pulleys, - - - - - - - - 170
Gear-Wheels, - - - - - - - 175
Fixed and Movable Pulleys, - - - - 183
The Inclined Plane, - - - - - 185
The Screw, - - - - - - - 187
Friction, - - - - - - - 189
Centrifugal Force, - - - - - 194
Specific Gravity, - - - - - - 196
Work and Energy, - - - - - 197
Belts, - - - - - - - - 201
Horsepower of Gears, - - - - - 204
Hydrostatics, - - - - - - 207
Buoyant Effects of Water, - - - - 219
Pneumatics, - - - - - - - 222
Pneumatic Machines, - - - - - 231
Pumps, - - - - - - - - 241
Strength of Materials, - - - - - 247
Tensile Strength of Materials, - - - 248
Chains, - - - - - - - - 251
Hemp Ropes, - - - - - - 252
Wire Ropes, - - - - - - - 253
Crushing Strength of Materials, - - - 255
Transverse Strength of Materials, - - 261
Shearing Strength of Materials, - - - 266
Line Shafting, - - - - - - 268

QUESTIONS AND EXAMPLES.
Arithmetic, - - - - - - - 275
Arithmetic, *continued*, - - - - - 283
Mensuration and Use of Letters in Algebraic Formulas, - - - - - - - 291
Mechanics, - - - - - - - 299
Mechanics, *continued*, - - - - - 309

ARITHMETIC.

DEFINITIONS.

1. Arithmetic is the art of reckoning, or the study of numbers.

2. A **unit** is *one*, or a single thing, as *one*, *one* bolt, *one* pulley, *one* dozen.

3. A **number** is a unit or a collection of units, as *one*, *three* engines, *five* boilers.

4. The **unit of a number** is one of the collection of units which constitutes the number. Thus, the unit of *twelve* is *one*, of *twenty* dollars is *one* dollar, of *one hundred* bolts is *one* bolt.

5. A **concrete number** is a number applied to some particular kind of object or quantity, as three *grate bars*, five *dollars*, ten *pounds*.

6. An **abstract number** is a number that is not applied to any object or quantity, as *three, five, ten*.

7. Like numbers are numbers which express units of the *same kind*, as 6 *days* and 10 *days*, 2 *feet* and 5 *feet*.

8. Unlike numbers are numbers which express units of *different kinds*, as ten *months* and eight *miles*, seven *wrenches* and five *bolts*.

NOTATION AND NUMERATION.

9. Numbers are expressed in three ways: (1) by words; (2) by figures; (3) by letters.

10. Notation is the art of expressing numbers by figures or letters.

11. Numeration is the art of reading the numbers which have been expressed by figures or letters.

ARITHMETIC.

12. The **Arabic notation** is the method of expressing numbers by figures. This method employs ten different **figures** to represent numbers, viz. :

Figures	0	1	2	3	4	5	6	7	8	9
Names	*naught, cipher or zero;*	*one*	*two*	*three*	*four*	*five*	*six*	*seven*	*eight*	*nine*

The first character (0) is called **naught, cipher, or zero**, and, when standing alone, has no value.

The other nine figures are called **digits**, and each one has a value of its own.

Any whole number is called an **integer**.

13. As there are only ten *figures* used in expressing numbers, each *figure* must express a different *value* at different times.

14. The value of a figure depends upon its *position* in relation to others.

15. Figures have **simple** values and **local** or **relative** values.

16. The **simple** value of a figure is the value it expresses when standing alone.

17. The **local** or **relative** value is the *increased* value it expresses by having other figures placed on its right.

For instance, if we see the figure 6 standing alone, thus. 6
we consider it as *six units*, or simply **six**.

Place another 6 to the *left* of it; thus. 66

The original figure is still *six units*, but the second one is *ten times* 6, or 6 **tens**.

If a third 6 be now placed still one place further to the *left*, it is increased in value *ten times* more, thus making it 6 **hundreds** 666

A fourth 6 would be 6 **thousands** 6666

A fifth 6 would be 6 **tens of thousands,** or **sixty thousand** 66666

A sixth 6 would be 6 **hundreds of thousands** . 666666

A seventh 6 would be 6 **millions** 6666666

The entire line of seven figures is read *six millions, six hundred sixty-six thousands, six hundred sixty-six.*

ARITHMETIC.

18. The **increased value** of each of these figures is its *local* or *relative* value. Each figure is *ten times* greater in value than the one immediately on its *right*.

19. The **cipher** (0) has no value itself, but it is useful in determining the place of other figures. To represent the number *four hundred five*, two digits only are necessary, one to represent *four hundred*, and the other to represent *five units;* but if these two digits are placed together, as 45, the 4 (being in the second place) will mean 4 *tens*. To mean 4 *hundreds*, the 4 should have two figures on its right, and a *cipher* is therefore inserted in the place usually given to *tens*, to show that the number is composed of *hundreds* and *units* only, and that there are no *tens*. Four hundred five is therefore expressed as 405. If the number were *four thousand and five*, two ciphers would be inserted; thus, 4005. If it were *four hundred fifty*, it would have the *cipher* at the right-hand side to show that there were no *units*, and only *hundreds and tens;* thus, 450. *Four thousand and fifty* would be expressed 4050, the first cipher indicating that there are no hundreds and the second that there are no units.

NOTE.—When speaking of the figures of a number by referring to them as first figure, second figure, etc., always begin to count at the *left*. Thus, in the number 41,625, 4 is the first figure, 6 the third figure, 5 the fifth or last figure, etc.

20. In *reading* figures, it is usual to point off the number into groups of three figures each, beginning with the right-hand or **units** column, a comma (,) being used to point off these groups.

Billions.			Millions.			Thousands.			Units.		
Hundreds of Billions.	Tens of Billions.	Billions.	Hundreds of Millions.	Tens of Millions.	Millions.	Hundreds of Thousands.	Tens of Thousands.	Thousands.	Hundreds of Units.	Tens of Units.	Units.
4	3	2,	1	9	8,	7	6	5,	4	3	2

In *pointing off* these figures, begin at the right-hand figure and count—*units, tens, hundreds;* the next group of three figures is *thousands,* therefore, we insert a comma (,) before beginning with them. Beginning at the figure 5, we say *thousands, tens of thousands, hundreds of thousands,* and insert another comma; we next read *millions, tens of millions, hundreds of millions,* and insert another comma; we then read *billions, tens of billions, hundreds of billions.*

The entire line of figures would be read: *Four hundred thirty-two billions, one hundred ninety-eight millions, seven hundred sixty-five thousands, four hundred thirty-two.* When we thus *read* a line of figures it is called **numeration,** and if the **numeration** be changed back to *figures,* it is called **notation.**

For instance, the writing of the figures,
$$72,584,623,$$
would be the **notation,** and the **numeration** would be *seventy-two millions, five hundred eighty-four thousands, six hundred twenty-three.*

21. NOTE.—It is customary to leave the *s* off the words millions, thousands, etc., in cases like the above, both in speaking and writing; hence, the above would usually be expressed, seventy-two million, five hundred eighty-four thousand, six hundred twenty-three.

22. The four fundamental processes of Arithmetic are **addition, subtraction, multiplication,** and **division.** They are called fundamental processes, because all operations in Arithmetic are based upon them.

ADDITION.

23. Addition is the *process* of *finding* the *sum* of *two* or *more* numbers. The sign of addition is $+$. It is read *plus,* and means *more.* Thus, $5+6$ is read 5 *plus* 6, and means that 5 and 6 are to be added.

24. The sign of equality is $=$. It is read *equals* or *is equal to.* Thus, $5+6=11$ may be read 5 plus 6 *equals* 11.

25. *Like numbers* can be added, but *unlike numbers* cannot. Thus, 6 dollars *can* be added to 7 dollars, and the *sum* will be 13 dollars, but 6 dollars *cannot* be added to 7 *feet.*

ARITHMETIC.

26. The following table gives the sum of any two numbers from 1 to 12:

TABLE 1.

1 and 1 are 2	2 and 1 are 3	3 and 1 are 4	4 and 1 are 5
1 and 2 are 3	2 and 2 are 4	3 and 2 are 5	4 and 2 are 6
1 and 3 are 4	2 and 3 are 5	3 and 3 are 6	4 and 3 are 7
1 and 4 are 5	2 and 4 are 6	3 and 4 are 7	4 and 4 are 8
1 and 5 are 6	2 and 5 are 7	3 and 5 are 8	4 and 5 are 9
1 and 6 are 7	2 and 6 are 8	3 and 6 are 9	4 and 6 are 10
1 and 7 are 8	2 and 7 are 9	3 and 7 are 10	4 and 7 are 11
1 and 8 are 9	2 and 8 are 10	3 and 8 are 11	4 and 8 are 12
1 and 9 are 10	2 and 9 are 11	3 and 9 are 12	4 and 9 are 13
1 and 10 are 11	2 and 10 are 12	3 and 10 are 13	4 and 10 are 14
1 and 11 are 12	2 and 11 are 13	3 and 11 are 14	4 and 11 are 15
1 and 12 are 13	2 and 12 are 14	3 and 12 are 15	4 and 12 are 16
5 and 1 are 6	6 and 1 are 7	7 and 1 are 8	8 and 1 are 9
5 and 2 are 7	6 and 2 are 8	7 and 2 are 9	8 and 2 are 10
5 and 3 are 8	6 and 3 are 9	7 and 3 are 10	8 and 3 are 11
5 and 4 are 9	6 and 4 are 10	7 and 4 are 11	8 and 4 are 12
5 and 5 are 10	6 and 5 are 11	7 and 5 are 12	8 and 5 are 13
5 and 6 are 11	6 and 6 are 12	7 and 6 are 13	8 and 6 are 14
5 and 7 are 12	6 and 7 are 13	7 and 7 are 14	8 and 7 are 15
5 and 8 are 13	6 and 8 are 14	7 and 8 are 15	8 and 8 are 16
5 and 9 are 14	6 and 9 are 15	7 and 9 are 16	8 and 9 are 17
5 and 10 are 15	6 and 10 are 16	7 and 10 are 17	8 and 10 are 18
5 and 11 are 16	6 and 11 are 17	7 and 11 are 18	8 and 11 are 19
5 and 12 are 17	6 and 12 are 18	7 and 12 are 19	8 and 12 are 20
9 and 1 are 10	10 and 1 are 11	11 and 1 are 12	12 and 1 are 13
9 and 2 are 11	10 and 2 are 12	11 and 2 are 13	12 and 2 are 14
9 and 3 are 12	10 and 3 are 13	11 and 3 are 14	12 and 3 are 15
9 and 4 are 13	10 and 4 are 14	11 and 4 are 15	12 and 4 are 16
9 and 5 are 14	10 and 5 are 15	11 and 5 are 16	12 and 5 are 17
9 and 6 are 15	10 and 6 are 16	11 and 6 are 17	12 and 6 are 18
9 and 7 are 16	10 and 7 are 17	11 and 7 are 18	12 and 7 are 19
9 and 8 are 17	10 and 8 are 18	11 and 8 are 19	12 and 8 are 20
9 and 9 are 18	10 and 9 are 19	11 and 9 are 20	12 and 9 are 21
9 and 10 are 19	10 and 10 are 20	11 and 10 are 21	12 and 10 are 22
9 and 11 are 20	10 and 11 are 21	11 and 11 are 22	12 and 11 are 23
9 and 12 are 21	10 and 12 are 22	11 and 12 are 23	12 and 12 are 24

This table should be carefully committed to memory. Since 0 has no value, the sum of any number and 0 is the number itself; thus, 17 and 0 are 17.

27. For *addition*, place the numbers to be added directly under each other, taking care to place *units* under *units*, *tens* under *tens*, *hundreds* under *hundreds*, and so on.

When the numbers are thus written, the *right-hand figure* of *one number* is placed *directly under the right-hand figure* of the *number above it*, thus bringing the unit figures of all the numbers to be added in the same vertical line. Proceed as in the following examples:

28. EXAMPLE.—What is the sum of 131, 222, 21, 2, and 413?
SOLUTION.—
$$\begin{array}{r} 131 \\ 222 \\ 21 \\ 2 \\ 413 \\ \hline sum\ \ 789\ \ Ans. \end{array}$$

EXPLANATION.—After placing the numbers in proper order, begin at the bottom of the right-hand or *units* column, and add, mentally repeating the different sums. Thus, three and two are five and one are six and two are eight and one are nine, the sum of the numbers in *units* column. Place the 9 directly beneath as the first or *units* figure in the sum.

The sum of the numbers in the next or *tens* column equals 8 *tens*, which is the second or *tens* figure in the sum.

The sum of the numbers in the next or *hundreds* column equals 7 *hundreds*, which is the third or *hundreds* figure in the sum.

The sum or answer is 789.

29. EXAMPLE.—What is the sum of 425, 36, 9,215, 4, and 907?
SOLUTION.—
$$\begin{array}{r} 425 \\ 36 \\ 9215 \\ 4 \\ 907 \\ \hline 27 \\ 60 \\ 1500 \\ 9000 \\ \hline sum\ \ 10587\ \ Ans. \end{array}$$

EXPLANATION.—The sum of the numbers in the first or units column is seven and four are eleven and five are sixteen and six are twenty-two and five are twenty-seven, or 27 units; i. e., two tens and seven units. Write 27 as shown.

ARITHMETIC. 7

The sum of the numbers in the second or tens column is six tens, or 60. Write 60 underneath 27 as shown. The sum of the numbers in the third or hundreds column is 15 hundreds, or 1,500. Write 1,500 under the two preceding results as shown. There is only one number in the fourth or thousands column, nine, which represents 9,000. Write 9,000 under the three preceding results. Adding these four results, the sum is 10,587, which is the sum of 425, 36, 9,215, 4, and 907.

NOTE.—It frequently happens, when adding a long column of figures, that the sum of two numbers, one of which does not occur in the addition table, is required. Thus, in the first column above, the sum of 16 and 6 was required. We know from the table that $6 + 6 = 12$; hence, the first figure of the sum is 2. Now, the sum of any number less than 20 and of any number less than 10 must be less than thirty, since $20 + 10 = 30$; therefore, the sum is 22. Consequently, in cases of this kind, add the first figure of the larger number to the smaller number and, if the result is greater than 9, increase the second figure of the larger number by 1. Thus, $44 + 7 = ?$ $4 + 7 = 11$; hence, $44 + 7 = 51$.

30. The addition may also be performed as follows:

$$\begin{array}{r} 425 \\ 36 \\ 9215 \\ 4 \\ 907 \\ \hline \text{sum } 10587 \text{ Ans.} \end{array}$$

EXPLANATION.—The sum of the numbers in *units* column $= 27$ *units*, or 2 *tens* and 7 *units*. Write the 7 *units* as the first or right-hand figure in the sum. Reserve the two *tens* and add them to the figures in *tens* column. The sum of the figures in the *tens* column, plus the 2 *tens* reserved and carried from the *units* column $= 8$, which is written down as the second figure in the sum. There is nothing to carry to the next column, because 8 is less than 10. The sum of the numbers in the next column is 15 *hundreds*, or 1 *thousand* and 5 *hundreds*. Write down the 5 as the third or *hundreds* figure in the sum and carry the 1 to the next column. $1 + 9 = 10$, which is written down at the left of the other figures.

The second method saves space and figures, but the first is to be preferred when adding a long column.

31. Example.—Add the numbers in the column below.

Solution.—

```
    890
     82
     90
    393
    281
     80
    770
     83
    492
     80
    383
     84
    191
sum 3899  Ans.
```

Explanation.—The sum of the digits in the first column equals 19 *units*, or 1 *ten* and 9 *units*. Write down the 9 and carry 1 to the next column. The sum of the digits in the second column $+1 = 109$ *tens*, or 10 *hundreds* and 9 *tens*. Write down the 9 and carry the 10 to the next column. The sum of the digits in this column plus the 10 reserved $= 38$. The entire sum is 3,899.

32. Rule 1.—(*a*) *Begin at the right, add each column separately, and write the sum, if it be only one figure, under the column added.*

(*b*) *If the sum of any column consists of two or more figures, put the right-hand figure of the sum under that column, and add the remaining figure or figures to the next column.*

33. Proof.—*To prove addition, add each column from top to bottom. If you obtain the same result as by adding from bottom to top, the work is probably correct.*

EXAMPLES FOR PRACTICE.

34. Find the sum of

(*a*) $104 + 203 + 613 + 214.$
(*b*) $1,875 + 3,143 + 5,826 + 10,832.$
(*c*) $4,865 + 2,145 + 8,173 + 40,084.$
(*d*) $14,204 + 8,173 + 1,065 + 10,042.$

Ans. { (*a*) 1,134.
(*b*) 21,676.
(*c*) 55,267.
(*d*) 33,484.

(e) 10,832 + 4,145 + 3,133 + 5,872.
(f) 214 + 1,231 + 141 + 5,000.
(g) 123 + 104 + 425 + 126 + 327.
(h) 6,354 + 2,145 + 2,042 + 1,111 + 3,333.

Ans. {
(e) 23,982.
(f) 6,586.
(g) 1,105.
(h) 14,985.
}

1. A week's record of coal burned in an engine room is as follows: Monday, 1,800 pounds; Tuesday, 1,655 pounds; Wednesday, 1,725 pounds; Thursday, 1,690 pounds; Friday, 1,648 pounds; Saturday, 1,020 pounds. How much coal was burned during the week?
Ans. 9,538 pounds.

2. A steam pump pumps out of a cistern in one hour 4,200 gallons; in the next hour, 5,420 gallons, and in 45 minutes more, an additional 3,600 gallons, when the cistern becomes empty. How many gallons were in the cistern at first? Ans. 13,220 gallons.

3. What is the total cost of a steam plant, the several items of expense being as follows: Steam engine, $900; boiler, $775; fittings and connections, $225; erecting the plant, $125; engine house, $650?
Ans. $2,675.

SUBTRACTION.

35. In Arithmetic, **subtraction** is the process of finding how much greater one number is than another.

The greater of the two numbers is called the **minuend**.
The smaller of the two numbers is called the **subtrahend**.
The number left after subtracting the *subtrahend* from the *minuend* is called the **difference** or **remainder**.

36. The sign of subtraction is —. It is read **minus**, and means *less*. Thus, 12 — 7 is read 12 *minus* 7, and means that 7 is to be taken from 12.

37. EXAMPLE.—From 7,568 take 3,425.
SOLUTION.—
    minuend 7 5 6 8
   subtrahend 3 4 2 5
   remainder 4 1 4 3 Ans.

EXPLANATION.—Begin at the right-hand or *units* column and subtract in succession each figure in the subtrahend from the one directly above it in the minuend, and write the remainders below the line. The result is the entire remainder.

38. When there are more figures in the *minuend* than in the *subtrahend*, and when some figures in the *minuend* are *less* than the figures directly under them in the *subtrahend*, proceed as in the following example:

EXAMPLE.—From 8,453 take 844.

SOLUTION.—
$$\begin{array}{r} \textit{minuend}\ \ 8453 \\ \textit{subtrahend}\ \ \ 844 \\ \hline \textit{remainder}\ \ 7609\ \text{Ans.} \end{array}$$

EXPLANATION.—Begin at the right-hand or *units* column to subtract. We can not take 4 from 3, and must, therefore, borrow 1 from 5 in *tens* column and annex it to the 3 in *units* column. The 1 *ten* = 10 *units*, which added to the 3 in the *units* column = 13 *units*. 4 from 13 = 9, the first or *units* figure in the remainder.

Since we borrowed 1 from the 5, only 4 remains; 4 from 4 = 0, the second or *tens* figure. We can not take 8 from 4, so borrow 1 *thousand* or 10 *hundreds* from 8; 10 *hundreds* + 4 *hundreds* = 14 *hundreds*.

8 from 14 = 6, the third or *hundreds* figure in the remainder.

Since we borrowed 1 from 8 only 7 remains, from which there is nothing to subtract; therefore, 7 is the next figure in the remainder or answer.

The operation of borrowing is placing 1 before the figure following the one from which it is borrowed. In the above example the one borrowed from 5 is placed before 3, making it 13, from which we subtract 4. The 1 borrowed from 8 is placed before 4, making 14, from which 8 is taken.

39. EXAMPLE.—Find the difference between 10,000 and 8,763.

SOLUTION.—
$$\begin{array}{r} \textit{minuend}\ \ 10000 \\ \textit{subtrahend}\ \ \ 8763 \\ \hline \textit{remainder}\ \ \ 1237\ \text{Ans.} \end{array}$$

EXPLANATION.—In the above example we borrow 1 from the second column and place it before 0, making 10; 3 from 10 = 7. In the same way we borrow 1 and place it before the next cipher, making 10; but as we have borrowed 1 from this column and taken it to the *units* column, only 9 remains, from which to subtract 6, 6 from 9 = 3. For the same reason we subtract 7 from 9 and 8 from 9 for the next two figures, and obtain a total remainder of 1,237.

40. Rule 2.—*Place the subtrahend or smaller number under the minuend or larger number, in the same manner as for addition, and proceed as in Arts.* **37, 38,** *and* **39**.

41. Proof.—*To prove an example in subtraction, add the subtrahend and remainder. The sum should equal the minuend. If it does not, a mistake has been made, and the work should be done over.*

Proof of the above example:

$$\begin{array}{rr} subtrahend & 8763 \\ remainder & 1237 \\ \hline minuend & 10000 \end{array}$$

EXAMPLES FOR PRACTICE.

42. From
(*a*) 94,278 take 62,574.
(*b*) 53,714 take 25,824.
(*c*) 71,832 take 58,109.
(*d*) 20,804 take 10,408.
(*e*) 310,465 take 102,141.
(*f*) (81,043 + 1,041) take 14,831.
(*g*) (20,482 + 18,216) take 21,214.
(*h*) (2,040 + 1,213 + 542) take 3,791.

Ans.
{
(*a*) 31,704.
(*b*) 27,890.
(*c*) 13,723.
(*d*) 10,396.
(*e*) 208,324.
(*f*) 67,253.
(*g*) 17,484.
(*h*) 4.
}

1. A cistern is fed by two pipes which supply 1,200 and 2,250 gallons per hour, respectively, and is being emptied by a pump which delivers 5,800 gallons per hour. Starting with 8,000 gallons in the cistern, how much water does it contain at the end of an hour? Ans. 5,650 gallons.

2. A train in running from New York to Buffalo travels 38 miles the first hour, 42 the second, 39 the third, 56 the fourth, 52 the fifth, and 48 the sixth hour. How many miles remain to be traveled at the end of the sixth hour, the distance between the two places being 410 miles? Ans. 135 miles.

3. On Monday morning a bank had on hand $2,862. During the day $1,831 were deposited and $2,172 drawn out; on Tuesday, $3,126 were deposited, and $1,954 drawn out. How many dollars were on hand Wednesday morning? Ans. $3,693.

MULTIPLICATION.

43. To **multiply** a number is to *add* it to itself a certain number of times.

44. **Multiplication** is the process of multiplying one number by another.

The *number* thus added to itself, or the number to be multiplied, is called the **multiplicand.**

The *number* which shows how many times the *multiplicand* is to be taken, or the *number* by which we *multiply*, is called the **multiplier.**

The result obtained by multiplying is called the **product.**

45. The sign of multiplication is ×. It is read *times* or *multiplied by.* Thus, 9 × 6 is read 9 *times* 6, or 9 *multiplied by* 6.

46. It matters not in what order the numbers to be multiplied together are placed. Thus, 6 × 9 is the same as 9 × 6.

47. In the following table, the product of any two numbers (neither of which exceeds twelve) may be found:

ARITHMETIC.

TABLE 2.

1 times	1 is	1	2 times	1 are	2	3 times	1 are	3
1 times	2 are	2	2 times	2 are	4	3 times	2 are	6
1 times	3 are	3	2 times	3 are	6	3 times	3 are	9
1 times	4 are	4	2 times	4 are	8	3 times	4 are	12
1 times	5 are	5	2 times	5 are	10	3 times	5 are	15
1 times	6 are	6	2 times	6 are	12	3 times	6 are	18
1 times	7 are	7	2 times	7 are	14	3 times	7 are	21
1 times	8 are	8	2 times	8 are	16	3 times	8 are	24
1 times	9 are	9	2 times	9 are	18	3 times	9 are	27
1 times	10 are	10	2 times	10 are	20	3 times	10 are	30
1 times	11 are	11	2 times	11 are	22	3 times	11 are	33
1 times	12 are	12	2 times	12 are	24	3 times	12 are	36
4 times	1 are	4	5 times	1 are	5	6 times	1 are	6
4 times	2 are	8	5 times	2 are	10	6 times	2 are	12
4 times	3 are	12	5 times	3 are	15	6 times	3 are	18
4 times	4 are	16	5 times	4 are	20	6 times	4 are	24
4 times	5 are	20	5 times	5 are	25	6 times	5 are	30
4 times	6 are	24	5 times	6 are	30	6 times	6 are	36
4 times	7 are	28	5 times	7 are	35	6 times	7 are	42
4 times	8 are	32	5 times	8 are	40	6 times	8 are	48
4 times	9 are	36	5 times	9 are	45	6 times	9 are	54
4 times	10 are	40	5 times	10 are	50	6 times	10 are	60
4 times	11 are	44	5 times	11 are	55	6 times	11 are	66
4 times	12 are	48	5 times	12 are	60	6 times	12 are	72
7 times	1 are	7	8 times	1 are	8	9 times	1 are	9
7 times	2 are	14	8 times	2 are	16	9 times	2 are	18
7 times	3 are	21	8 times	3 are	24	9 times	3 are	27
7 times	4 are	28	8 times	4 are	32	9 times	4 are	36
7 times	5 are	35	8 times	5 are	40	9 times	5 are	45
7 times	6 are	42	8 times	6 are	48	9 times	6 are	54
7 times	7 are	49	8 times	7 are	56	9 times	7 are	63
7 times	8 are	56	8 times	8 are	64	9 times	8 are	72
7 times	9 are	63	8 times	9 are	72	9 times	9 are	81
7 times	10 are	70	8 times	10 are	80	9 times	10 are	90
7 times	11 are	77	8 times	11 are	88	9 times	11 are	99
7 times	12 are	84	8 times	12 are	96	9 times	12 are	108
10 times	1 are	10	11 times	1 are	11	12 times	1 are	12
10 times	2 are	20	11 times	2 are	22	12 times	2 are	24
10 times	3 are	30	11 times	3 are	33	12 times	3 are	36
10 times	4 are	40	11 times	4 are	44	12 times	4 are	48
10 times	5 are	50	11 times	5 are	55	12 times	5 are	60
10 times	6 are	60	11 times	6 are	66	12 times	6 are	72
10 times	7 are	70	11 times	7 are	77	12 times	7 are	84
10 times	8 are	80	11 times	8 are	88	12 times	8 are	96
10 times	9 are	90	11 times	9 are	99	12 times	9 are	108
10 times	10 are	100	11 times	10 are	110	12 times	10 are	120
10 times	11 are	110	11 times	11 are	121	12 times	11 are	132
10 times	12 are	120	11 times	12 are	132	12 times	12 are	144

This table should be carefully committed to memory.
Since 0 has no value, the product of 0 and any number is 0.

ARITHMETIC.

48. To multiply a number by one figure only:

EXAMPLE.—Multiply 425 by 5.

SOLUTION.—
$$\begin{array}{r} \textit{multiplicand} \quad 425 \\ \textit{multiplier} \quad 5 \\ \hline \textit{product} \quad 2125 \quad \text{Ans.} \end{array}$$

EXPLANATION.—For convenience, the *multiplier* is generally written *under* the *right-hand figure* of the *multiplicand*. On looking in the multiplication table, we see that 5 × 5 are 25. *Multiplying* the *first figure* at the *right* of the *multiplicand*, or 5, by the *multiplier* 5, it is seen that 5 times 5 units are 25 units, or 2 tens and 5 units. Write the 5 *units* in *units place* in the *product*, and *reserve* the 2 tens to *add* to the *product* of *tens*. Looking in the multiplication table again, we see that 5 × 2 are 10. *Multiplying* the *second figure* of the *multiplicand* by the *multiplier* 5, we see *that* 5 *times* 2 tens are 10 tens, *plus* the 2 tens *reserved*, are 12 tens, or 1 hundred plus 2 tens. Write the *2 tens* in tens place, and *reserve* the 1 hundred to *add* to the product of *hundreds*. Again, we see by the multiplication table that 5 × 4 are 20. *Multiplying* the *third* or *last figure* of the *multiplicand* by the *multiplier* 5, we see that 5 times 4 hundreds are 20 hundreds, *plus* the 1 hundred *reserved*, are 21 hundreds, or 2 thousands *plus* 1 hundred, which we write in *thousands* and *hundreds places*, respectively.

Hence, the *product* is 2,125.

This *result* is the same as adding 425 five times. Thus,

$$\begin{array}{r} 425 \\ 425 \\ 425 \\ 425 \\ 425 \\ \hline \textit{sum} \quad 2125 \quad \text{Ans.} \end{array}$$

EXAMPLES FOR PRACTICE.

49. Find the product of

(a) 61,483 × 6.
(b) 12,375 × 5.
(c) 10,426 × 7.
(d) 10,835 × 8.

Ans. {
(a) 368,898.
(b) 61,875.
(c) 72,982.
(d) 32,505.
}

ARITHMETIC.

(e) 98,376 × 4.
(f) 10,873 × 8.
(g) 71,543 × 9.
(h) 218,734 × 2.

Ans. {
(e) 393,504.
(f) 86,984.
(g) 643,887.
(h) 437,468.
}

1. A stationary engine makes 5,520 revolutions per hour. Running 9 hours a day, 5 days in the week, and 5 hours on Saturday, how many revolutions would it make in 4 weeks? Ans. 1,104,000 revolutions.

2. An engineer earns $650 a year, and his average expenses are $548. How much could he save in 8 years at that rate? Ans. $816.

3. The connection between an engine and boiler is made up of 5 lengths of pipe, three of which are 12 feet long, one 2 feet 6 inches long, and one 8 feet 6 inches long. If the pipe weighs 9 pounds per foot, what is the total weight of the pipe used? Ans. 423 pounds.

50. To multiply a number by two or more figures:

EXAMPLE.—Multiply 475 by 234.

SOLUTION.— *multiplicand*   475
             multiplier   234
                               ─────
                               1900
                               1425
                                950
                              ──────
                product 111150 Ans.

EXPLANATION.—For convenience, the *multiplier* is generally written *under* the *multiplicand*, placing units under units, tens under tens, etc.

We *can not* multiply by 234 at one operation; we must, therefore, *multiply* by the *parts* and then *add* the **partial products.**

The parts by which we are to multiply are 4 units, 3 tens, and 2 hundreds. 4 times 475 = 1,900, the *first partial product;* 3 times 475 = 1,425, the *second partial product*, the *right-hand figure* of which is *written directly under* the *figure multiplied by*, or 3; 2 times 475 = 950, the *third partial product*, the *right-hand figure* of which is *written directly under* the *figure multiplied by*, or 2.

The sum of these *three partial products* is 111,150, which is the *entire product*.

ARITHMETIC.

51. Rule 3.—(*a*) *Write the multiplier under the multiplicand, so that units are under units, tens under tens, etc.*

(*b*) *Begin at the right and multiply each figure of the multiplicand by each successive figure of the multiplier, placing the right-hand figure of each partial product directly under the figure used as a multiplier.*

(*c*) *The sum of the partial products will equal the required product.*

52. Proof.—*Review the work carefully, or multiply the multiplier by the multiplicand; if the results agree, the work is correct.*

53. When there is a *cipher* in the *multiplier*, multiply the entire multiplicand by it; since the result will be zero, place a cipher under the cipher in the multiplier. Thus,

```
   (a)         (b)          (c)           (d)
    0           2           15            708
   ×0          ×0           × 0           ×  0
   ──          ──           ──            ────
    0  Ans.     0  Ans.      0  Ans.        0  Ans.

   (e)              (f)                (g)
   3114            4008              31264
    203             305               1002
   ────            ─────             ──────
   9342           20040              62528
  62280          120240            3126400
  ──────         ──────            ────────
  632142 Ans.   1222440 Ans.      31326528 Ans.
```

In examples (*e*), (*f*), and (*g*), we multiply by 0 as directed above; then multiply by the next figure of the multiplier and place the first figure of the product alongside the 0, as shown.

EXAMPLES FOR PRACTICE.

54. Find the product of

(*a*) 3,842 × 26.
(*b*) 3,716 × 45.
(*c*) 1,817 × 124.
(*d*) 675 × 38.

Ans. { (*a*) 99,892.
(*b*) 167,220.
(*c*) 225,308.
(*d*) 25,650.

ARITHMETIC. 17

(e) 1,875 × 33.
(f) 4,836 × 47.
(g) 5,682 × 543.
(h) 3,257 × 246.
(i) 2,875 × 302.
(j) 17,819 × 1,004.
(k) 38,674 × 205.
(l) 18,304 × 100.
(m) 7,834 × 10.
(n) 87,543 × 1,000.
(o) 48,763 × 100.

Ans.
(e) 61,875.
(f) 227,292.
(g) 3,085,326.
(h) 801,222.
(i) 868,250.
(j) 17,890,276.
(k) 7,928,170.
(l) 1,830,400.
(m) 78,340.
(n) 87,543,000.
(o) 4,876,300.

1. If the area of a steam-engine piston is 113 square inches, what is the total pressure upon it when the steam pressure is 85 pounds per square inch? Ans. 9,605 pounds.

2. A steam engine, which indicated 164 horsepower, was found to consume 4 pounds of coal per horsepower per hour. Being replaced by a new engine, which was of the same horsepower as the other, another test was made, which showed a consumption of 3 pounds per horsepower per hour. What was the saving of coal for a year of 309 days, if the engine averaged to run 14 hours a day? Ans. 709,464 pounds.

3. Two steamers are 7,846 miles apart, and are sailing towards each other, one at the rate of 18 miles an hour, and the other at the rate of 15 miles an hour. How far apart will they be at the end of 205 hours? Ans. 1,081 miles.

DIVISION.

55. Division is the process of finding how many times one number is contained in another of the same kind.

The number to be *divided* is called the **dividend.**

The number by which we *divide* is called the **divisor.**

The number which *shows* how many times the *divisor* is *contained* in the *dividend* is called the **quotient.**

56. The sign of division is ÷. It is read *divided by*. 54 ÷ 9 is read 54 *divided by* 9. Another way to write 54 *divided by* 9 is $\frac{54}{9}$. Thus, $54 \div 9 = 6$, or $\frac{54}{9} = 6$.

In both of these cases 54 is the *dividend* and 9 is the *divisor.*

Division is the *reverse* of **multiplication.**

57. To divide when the divisor consists of but one figure, proceed as in the following example:

EXAMPLE.—What is the quotient of 875 ÷ 7?

$$\begin{array}{r} \text{\textit{divisor dividend quotient}} \\ \text{SOLUTION.—} \quad 7)8\,7\,5(1\,2\,5 \quad \text{Ans.} \\ \underline{7} \\ 17 \\ \underline{14} \\ 35 \\ \underline{35} \\ \text{\textit{remainder}} \quad 0 \end{array}$$

EXPLANATION.—7 is contained in 8 *hundreds* 1 *hundred* times. Place the one as the first, or left-hand, figure of the quotient. Multiply the divisor 7 by the 1 *hundred* of the quotient, and place the product 7 *hundreds* under the 8 *hundreds* in the dividend, and subtract. Beside the remainder 1, bring down the 7 *tens*, making 17 *tens;* 17 divided by 7 = 2 times. Write the two as the second figure of the quotient. Multiply the divisor 7 by the 2, and subtract the product from 17. Beside the remainder 3, bring down the 5 *units* of the dividend, making 35 *units*. 7 is contained in 35, 5 times, which is placed in the quotient. Multiplying the divisor by the last figure of the quotient, 5 times 7 = 35, which subtracted from 35, under which it is placed, leaves 0. Therefore, the quotient is 125. This method is called **long division.**

58. In **short division,** only the divisor, dividend, and quotient are written, the operations being performed mentally.

$$\begin{array}{r} \text{\textit{dividend}} \\ \text{\textit{divisor}} \quad 7\,)\,8^1\,7^3\,5 \\ \text{\textit{quotient}} \quad 1\,\ 2\,\ 5 \quad \text{Ans.} \end{array}$$

The mental operation is as follows: 7 is contained in 8, once and one remainder; 1 placed before 7 makes 17; 7 is contained in 17, 2 times and 3 over; the 3 placed before 5 makes 35; 7 is contained in 35, 5 times. These partial quotients placed in order as they are found, make the entire quotient 125.

ARITHMETIC. 19

The small figures are placed in the example given to better illustrate the explanation; they are never written when actually performing division in this way.

59. If the *divisor* consists of *2 or more* figures, proceed as in the following example:

EXAMPLE.—Divide 2,702,826 by 63.

SOLUTION.—
```
        divisor   dividend    quotient
         63 ) 2 7 0 2 8 2 6 ( 4 2 9 0 2   Ans.
              2 5 2
              ─────
                1 8 2
                1 2 6
                ─────
                  5 6 8
                  5 6 7
                  ─────
                    1 2 6
                    1 2 6
                    ─────
                        0
```

EXPLANATION.—As 63 is not contained in the first two figures, 27, we must use the first three figures, 270. Now, by trial, we must find how many times 63 is contained in 270; 6 is contained in the first two figures of 270, 4 times. Place the 4 as the first or left-hand figure in the quotient. Multiply the divisor 63 by 4, and subtract the product 252 from 270. The remainder is 18, beside which we write the next figure of the dividend, 2, making 182. Now, 6 is contained in the first two figures of 182, 3 times, but on multiplying 63 by 3, we see that the product 189 is too great, so we try 2 as the second figure of the quotient. Multiplying the divisor 63 by 2, and subtracting the product 126 from 182, the remainder is 56, beside which we bring down the next figure of the dividend, making 568; 6 is contained in 56 about 9 times. Multiply the divisor 63 by 9 and subtract the product 567 from 568. The remainder is 1, and bringing down the next figure of the dividend, 2, gives 12. As 12 is smaller than 63, we write 0 in the quotient and bring down the next figure, 6, making 126. 63 is contained in 126, 2 times, without a remainder. Therefore, 42,902 is the quotient.

ARITHMETIC.

60. Rule 4.—(*a*) *Write the divisor at the left of the dividend, with a line between them.*

(*b*) *Find how many times the divisor is contained in the lowest number of the left-hand figures of the dividend that will contain it, and write the result at the right of the dividend, with a line between, for the first figure of the quotient.*

(*c*) *Multiply the divisor by this quotient; write the product under the partial dividend used, and subtract, annexing to the remainder the next figure of the dividend. Divide as before, and thus continue until all the figures of the dividend have been used.*

(*d*) *If any partial dividend will not contain the divisor, write a cipher in the quotient, annex the next figure of the dividend and proceed as before.*

(*e*) *If there be a remainder at last, write it after the quotient, with the divisor underneath.*

61. Proof.—*Multiply the quotient by the divisor, and add the remainder, if there be any, to the product. The result will be the dividend.*

$$
\begin{array}{r}
\text{divisor} \quad \text{dividend} \quad \text{quotient} \\
63\,)\,4235\,(\,67\tfrac{14}{63} \quad \text{Ans.} \\
378 \quad\quad\quad\quad \\ \hline
455 \\
441 \\ \hline
\text{remainder} \quad 14
\end{array}
$$

$$
\begin{array}{rr}
\text{Proof.} & \\
\text{quotient} & 67 \\
\text{divisor} & 63 \\ \hline
 & 201 \\
 & 402 \\ \hline
 & 4221 \\
\text{remainder} & 14 \\ \hline
\text{dividend} & 4235
\end{array}
$$

ARITHMETIC. 21

EXAMPLES FOR PRACTICE.

62. Divide the following:

(*a*) 126,498 by 58.		(*a*) 2,181.
(*b*) 3,207,594 by 767.		(*b*) 4,182.
(*c*) 11,408,202 by 234.		(*c*) 48,753.
(*d*) 2,100,315 by 581.	Ans.	(*d*) 3,615.
(*e*) 969,936 by 4,008.		(*e*) 242.
(*f*) 7,481,888 by 1,021.		(*f*) 7,328.
(*g*) 1,525,915 by 5,003.		(*g*) 305.
(*h*) 1,646,301 by 381.		(*h*) 4,321.

1. In a mile there are 5,280 feet. How many rails would it take to lay a double row one mile long, each rail being 30 feet long? Ans. 352 rails.

2. How many rivets will be required for the longitudinal seams of a cylindrical boiler 20 feet long, the joint being double riveted, and the rivets being spaced 4 inches apart? Ans. 120 rivets.

NOTE.—First find the length of the boiler in inches.

3. It requires 7,020,000 bricks to build a large foundry. How many teams will it require to draw the bricks in 60 days, if each team draws 6 loads per day and 1,500 bricks at a load? Ans. 13 teams.

NOTE.—Find how many loads 7,020,000 bricks make; then, how many days it will take one team to draw the brick.

CANCELATION.

63. Cancelation is the process of shortening operations in division by casting out equal factors from both dividend and divisor.

64. The **factors** of a number are *those numbers* which, when *multiplied* together, will *equal that number*. Thus, 5 and 3 are factors of 15, since $5 \times 3 = 15$. Likewise, 8 and 7 are the factors of 56, since $8 \times 7 = 56$.

65. A **prime number** is one which can not be divided by any number except itself and 1. Thus, 2, 3, 11, 29, etc., are prime numbers.

66. A **prime factor** is any factor that is a prime number.

Any number that is not a prime is called a **composite** number, and may be produced by multiplying together its prime factors. Thus, 60 is a composite number, and is equal to the product of its prime factors, $2 \times 2 \times 3 \times 5$.

Numbers are said to be **prime to each other** when no two of them can be divided by any number except 1; the numbers themselves *may* be either prime or composite. Thus, the numbers 3, 5, and 11 are prime to each other, so also are 22, 25, and 21 — all composite numbers.

67. Canceling *equal factors* from *both dividend and divisor* does *not* change the *quotient.*

The *canceling* of a *factor* in *both dividend and divisor* is the *same* as *dividing them both* by the *same number*, which, by the principle of division, does not *change the quotient.*

Write the *numbers* which make the *dividend* above the *line*, and those which make the *divisor* below it.

68. EXAMPLE.—Divide $4 \times 45 \times 60$ by 9×24.

SOLUTION.—Placing the dividend over the divisor, and canceling

$$\frac{\overset{}{4} \times \overset{5}{45} \times \overset{10}{60}}{\underset{1}{9} \times \underset{6}{24}} = \frac{50}{1} = 50. \text{ Ans.}$$

EXPLANATION.—The 4 in the dividend and 24 in the divisor are both divisible by 4, since 4 divided by 4 equals 1, and 24 divided by 4 equals 6. Cross off the four and write the 1 over it; also, cross off the 24 and write the 6 under it. Thus,

$$\frac{\overset{1}{4} \times 45 \times 60}{9 \times \underset{6}{24}}$$

60 in the dividend and 6 in the divisor are divisible by 6, since 60 divided by 6 equals 10, and 6 divided by 6 equals 1. Cross off the 60 and write 10 over it; also, cross off the 6 and write 1 under it. Thus,

$$\frac{\overset{1}{4} \times 45 \times \overset{10}{60}}{9 \times \underset{\underset{1}{6}}{24}}$$

Again, 45 in the dividend and 9 in the divisor are divisible by 9, since 45 divided by 9 equals 5, and 9 divided by 9 equals 1. Cross off the 45 and write the 5 over it; also, cross off the 9 and write the 1 under it. Thus,

$$\frac{\overset{1}{4} \times \overset{5}{45} \times \overset{10}{60}}{\underset{1}{9} \times \underset{\underset{1}{6}}{24}}$$

ARITHMETIC.

Since there are no two remaining numbers (one in the dividend and one in the divisor) divisible by any number except 1, without a remainder, it is impossible to cancel further.

Multiply all the uncanceled numbers in the dividend together, and divide their product by the product of all the uncanceled numbers in the divisor. The result will be the quotient. The product of all the uncanceled numbers in the dividend equals $5 \times 1 \times 10 = 50$; the product of all the uncanceled numbers in the divisor equals $1 \times 1 = 1$.

Hence, $\dfrac{\overset{1}{\cancel{4}} \times \overset{5}{\cancel{45}} \times \overset{10}{\cancel{90}}}{\underset{1}{\cancel{9}} \times \underset{\cancel{6}}{\cancel{24}}} = \dfrac{1 \times 5 \times 10}{1 \times 1} = 50.$ Ans.

It is usual to omit the 1's when canceling them, instead of writing them as above.

69. Rule 5.—(*a*) *Cancel the common factors from both the dividend and divisor.*

(*b*) *Then divide the product of the remaining factors of the dividend by the product of the remaining factors of the divisor, and the result will be the quotient.*

EXAMPLES FOR PRACTICE.

70. Divide

(*a*) $14 \times 18 \times 16 \times 40$ by $7 \times 8 \times 6 \times 5 \times 3$.
(*b*) $3 \times 65 \times 50 \times 100 \times 60$ by $30 \times 60 \times 13 \times 10$.
(*c*) $8 \times 4 \times 3 \times 9 \times 11$ by $11 \times 9 \times 4 \times 3 \times 8$.
(*d*) $164 \times 321 \times 6 \times 7 \times 4$ by $82 \times 321 \times 7$.
(*e*) $50 \times 100 \times 200 \times 72$ by $1,000 \times 144 \times 100$.
(*f*) $48 \times 63 \times 55 \times 49$ by $7 \times 21 \times 11 \times 48$.
(*g*) $110 \times 150 \times 84 \times 32$ by $11 \times 15 \times 100 \times 64$.
(*h*) $115 \times 120 \times 400 \times 1,000$ by $23 \times 1,000 \times 60 \times 800$.

Ans. $\begin{cases} (a)\ 32. \\ (b)\ 250. \\ (c)\ 1. \\ (d)\ 48. \\ (e)\ 5. \\ (f)\ 105. \\ (g)\ 42. \\ (h)\ 5. \end{cases}$

FRACTIONS.

71. A **fraction** is a *part* of a *whole number*: *One-half, one-third, two-fifths* are fractions.

72. *Two* numbers are required to express a fraction, one called the **numerator,** and the other the **denominator.**

73. The *numerator* is placed above the *denominator*, with a *line* between them; as, $\frac{2}{3}$. 3 is the *denominator*, and shows into how many *equal parts* the *unit* or *one* is divided. The *numerator* 2 shows how many of these *equal parts* are taken or considered. The *denominator* also indicates the *names* of the parts.

$\frac{1}{2}$ is read one-half.
$\frac{3}{4}$ is read three-fourths.
$\frac{3}{8}$ is read three-eighths.
$\frac{5}{16}$ is read five-sixteenths.
$\frac{29}{47}$ is read twenty-nine-forty-sevenths.

In the expression "$\frac{3}{4}$ of an apple," the *denominator* 4 shows that the apple is to be (or has been) cut into 4 *equal parts*, and the *numerator* 3 shows that *three of these parts*, or *fourths*, are taken or considered.

If each of the *parts*, or *fourths*, of the apple were cut in *two equal pieces*, there would then be twice as many pieces as before, or $4 \times 2 = 8$ pieces in all; one of these pieces would be called one-eighth, and would be expressed in figures as $\frac{1}{8}$. Three of these pieces would be called three-eighths, and written $\frac{3}{8}$. The words three-fourths, three-eighths, five-sixteenths, etc., are abbreviations of three one-fourths, three one-eighths, five one-sixteenths, etc. It is evident that the larger the *denominator*, the greater is the number of parts into which anything is divided; consequently, the parts themselves are smaller, and the value of the fraction is less for the same number of parts taken. In other words, $\frac{1}{9}$, for example, is smaller than $\frac{1}{8}$, because if an object be divided into 9 parts, the parts are smaller than if the same object had been divided into 8 parts; and, since $\frac{1}{9}$ is smaller than $\frac{1}{8}$, it is clear that 7 one-ninths is a smaller amount than 7 one-eighths. Hence, also, $\frac{7}{9}$ is less than $\frac{7}{8}$.

74. The **value** of a fraction is the result obtained by dividing the *numerator* by the *denominator;* as, $\frac{4}{2} = 2, \frac{6}{2} = 3$.

75. The line between the *numerator* and *denominator* means *divided by*, or $\div$.

$\frac{3}{4}$ is equivalent to $3 \div 4$.
$\frac{5}{8}$ is equivalent to $5 \div 8$.

ARITHMETIC.

76. The *numerator* and *denominator* of a fraction are called the **terms** of a fraction.

77. The *value* of a fraction whose *numerator* and *denominator* are equal, is 1.

$\frac{4}{4}$, or four-fourths, $=1$.
$\frac{8}{8}$, or eight-eighths, $= 1$.
$\frac{64}{64}$, or sixty-four sixty-fourths, $= 1$.

78. A **proper fraction** is a fraction whose *numerator* is *less* than its *denominator*. Its *value* is *less* than 1; as, $\frac{3}{4}$, $\frac{5}{8}$, $\frac{1}{16}$.

79. An **improper fraction** is a fraction whose *numerator equals* or is *greater than* the *denominator*. Its *value* is *one* or *more* than *one;* as, $\frac{4}{4}$, $\frac{8}{8}$, $\frac{16}{8}$.

80. A **mixed number** is a *whole* number and a *fraction* united. $4\frac{2}{3}$ is a mixed number, and is equivalent to $4 + \frac{2}{3}$. It is read *four and two-thirds.*

REDUCTION OF FRACTIONS.

81. **Reduction of fractions** is the process of changing their form without changing their *value.*

82. A *fraction* is reduced to *higher terms* by *multiplying both terms* of the *fraction* by the *same number.* Thus, $\frac{3}{4}$ is reduced to $\frac{6}{8}$ by multiplying both terms by 2.

$$\frac{3 \times 2}{4 \times 2} = \frac{6}{8}.$$

The *value* is not changed, since $\frac{3}{4} = \frac{6}{8}$. For, suppose that an object, say an apple, is divided into 8 equal parts. If these parts be arranged into 4 piles, each containing 2 parts, it is evident each pile will be composed of the same amount of the entire apple as would have been the case had the apple been originally cut into 4 equal parts. Now, if one of these piles (containing 2 parts) be removed, there will be 3 piles left each containing 2 equal parts, or 6 equal parts in all, i. e., six-eighths. But, since one pile, or one quarter, was removed, there are three-quarters left. Hence, $\frac{3}{4} = \frac{6}{8}$.

ARITHMETIC.

The same course of reasoning may be applied to any similar case. Therefore, multiplying both terms of a fraction by the same number does not alter its value.

$$\frac{3 \times 2}{4 \times 2} = \frac{6}{8}.$$

83. A *fraction* is reduced to *lower terms* by *dividing both terms* by the *same number*. Thus, $\frac{8}{10}$ is reduced to $\frac{4}{5}$ by dividing both terms by 2.

$$\frac{8 \div 2}{10 \div 2} = \frac{4}{5}.$$

That $\frac{8}{10} = \frac{4}{5}$ is readily seen from the explanation given in Art. **82**; for, multiplying both terms of the fraction $\frac{4}{5}$ by 2, $\frac{4 \times 2}{5 \times 2} = \frac{8}{10}$, and, if $\frac{4}{5} = \frac{8}{10}$, $\frac{8}{10}$ must equal $\frac{4}{5}$. Hence, dividing both terms of a fraction by the same number does not alter its value.

84. A *fraction* is reduced to its *lowest terms* when its *numerator and denominator* can not be *divided* by the *same number*, as $\frac{3}{4}$, $\frac{2}{3}$, $\frac{11}{14}$.

85. To reduce a whole number or mixed number to an improper fraction:

EXAMPLE.—How many *fourths* in 5?

SOLUTION.—Since there are 4 *fourths* in 1 ($\frac{4}{4} = 1$), in 5 there will be 5 × 4 fourths, or 20 fourths, $5 \times \frac{4}{4} = \frac{20}{4}$. Ans.

EXAMPLE.—Reduce $8\frac{3}{4}$ to an improper fraction.

SOLUTION.—$8 \times \frac{4}{4} = \frac{32}{4}$. $\frac{32}{4} + \frac{3}{4} = \frac{35}{4}$. Ans.

86. Rule 6.—*Multiply the whole number by the denominator of the fraction, add the numerator to the product, and place the denominator under the result.*

EXAMPLES FOR PRACTICE.

87. Reduce to improper fractions:

(a) $4\frac{1}{2}$.
(b) $5\frac{1}{3}$.
(c) $10\frac{8}{10}$.
(d) $37\frac{1}{4}$.
(e) $50\frac{3}{4}$.

Ans. $\begin{cases} (a) \ \frac{9}{2}. \\ (b) \ \frac{16}{3}. \\ (c) \ \frac{108}{10}. \\ (d) \ \frac{149}{4}. \\ (e) \ \frac{203}{4}. \end{cases}$

ARITHMETIC. 27

88. To reduce an improper fraction to a whole or mixed number:

EXAMPLE.—Reduce $\frac{21}{4}$ to a mixed number.

SOLUTION.—4 is contained in 21, 5 times and 1 remaining; as this is also divided by 4, its value is $\frac{1}{4}$. Therefore, $5 + \frac{1}{4}$, or $5\frac{1}{4}$, is the number. Ans.

89. Rule 7.—*Divide the numerator by the denominator; the quotient will be the whole number; the remainder, if there be any, will be the numerator of the fractional part of which the denominator is the same as the denominator of the improper fraction.*

EXAMPLES FOR PRACTICE.

90. Reduce to whole or mixed numbers:
(a) $\frac{121}{5}$.
(b) $\frac{185}{3}$.
(c) $\frac{701}{6}$.
(d) $\frac{442}{9}$.
(e) $\frac{36}{9}$.
(f) $\frac{115}{23}$.

Ans.
(a) $24\frac{1}{5}$.
(b) $61\frac{2}{3}$.
(c) $116\frac{5}{6}$.
(d) $49\frac{1}{9}$.
(e) 4.
(f) 5.

91. A common denominator of *two* or *more fractions* is a number which will contain all of the *denominators* of the *fractions* without a remainder. The **least common denominator** is the least number that will contain all of the denominators of the fractions without a remainder.

92. To find the least common denominator:

EXAMPLE.—Find the least common denominator of $\frac{1}{4}$, $\frac{1}{3}$, $\frac{1}{9}$, and $\frac{1}{16}$.

SOLUTION.—We first place the denominators in a row, separated by commas.

$$\begin{array}{r|l}
2 & 4, 3, 9, 16 \\ \hline
2 & 2, 3, 9, 8 \\ \hline
3 & 1, 3, 9, 4 \\ \hline
3 & 1, 1, 3, 4 \\ \hline
4 & 1, 1, 1, 4 \\ \hline
& 1, 1, 1, 1
\end{array}$$

$2 \times 2 \times 3 \times 3 \times 4 = 144$, the least common denominator. Ans.

EXPLANATION.—Divide each of them by some prime number which will divide at least two of them without a remainder (if possible), bringing down those denominators to the row below which will not contain the divisor without a remainder. Dividing each of the numbers by 2, the second row becomes 2, 3, 9, 8, since 2 will not divide 3 and 9 without a remainder. Dividing again by 2, the result is 1, 3, 9, 4. Dividing the third row by 3, the result is 1, 1, 3, 4. So continue until the last row contains only 1's. The product of all the divisors, or $2 \times 2 \times 3 \times 3 \times 4 = 144$, is the least common denominator.

93. EXAMPLE.—Find the least common denominator of $\frac{1}{9}, \frac{5}{12}, \frac{7}{18}$.

SOLUTION.—
$$\begin{array}{r|l} 3 & 9, 12, 18 \\ 3 & 3, 4, 6 \\ 2 & 1, 4, 2 \\ 2 & 1, 2, 1 \\ & 1, 1, 1 \end{array}$$

$3 \times 3 \times 2 \times 2 = 36$. Ans.

94. **To reduce two or more fractions to fractions having a common denominator:**

EXAMPLE.—Reduce $\frac{2}{3}, \frac{3}{4}$, and $\frac{1}{2}$ to fractions having a common denominator.

SOLUTION.—The common denominator is a number which will contain 3, 4, and 2. The least common denominator is 12, because it is the smallest number which can be divided by 3, 4, and 2 without a remainder.

$$\tfrac{2}{3} = \tfrac{8}{12}, \ \tfrac{3}{4} = \tfrac{9}{12}, \ \tfrac{1}{2} = \tfrac{6}{12}.$$

Reducing $\frac{2}{3}$, 3 is contained in 12, 4 times. By multiplying both numerator and denominator of $\frac{2}{3}$ by 4, we find

$$\frac{2 \times 4}{3 \times 4} = \frac{8}{12}.$$ In the same way we find $\tfrac{3}{4} = \tfrac{9}{12}$ and $\tfrac{1}{2} = \tfrac{6}{12}$.

95. **Rule 8.**—*Divide the common denominator by the denominator of the given fraction, and multiply both terms of the fraction by the quotient.*

EXAMPLES FOR PRACTICE.

96. Reduce to fractions having a common denominator:

(a) $\frac{2}{3}, \frac{3}{4}, \frac{1}{2}$.
(b) $\frac{3}{10}, \frac{1}{2}, \frac{7}{15}$.
(c) $\frac{7}{8}, \frac{7}{10}, \frac{11}{12}$.
(d) $\frac{2}{3}, \frac{3}{4}, \frac{11}{15}$.
(e) $\frac{1}{10}, \frac{3}{20}, \frac{9}{30}$.
(f) $\frac{7}{12}, \frac{13}{15}, \frac{11}{20}$.

Ans. $\begin{cases} (a) \ \frac{8}{12}, \frac{9}{12}, \frac{6}{12}. \\ (b) \ \frac{9}{30}, \frac{15}{30}, \frac{14}{30}. \\ (c) \ \frac{77}{80}, \frac{78}{80}, \frac{80}{80}. \\ (d) \ \frac{40}{60}, \frac{45}{60}, \frac{44}{60}. \\ (e) \ \frac{18}{60}, \frac{9}{60}, \frac{18}{60}. \\ (f) \ \frac{14}{15}, \frac{17}{15}, \frac{13}{15}. \end{cases}$

ARITHMETIC.

ADDITION OF FRACTIONS.

97. *Fractions can not be added unless they have a common denominator.* We can not add $\frac{3}{4}$ to $\frac{1}{8}$ as they now stand, since the denominators represent parts of different sizes. Fourths can not be added to eighths.

Suppose we divide an apple into 4 equal parts, and then divide 2 of these parts into two equal parts. It is evident that we shall have 2 one-fourths and 4 one-eighths. Now if we add these parts the result is $2 + 4 = 6$ something. But what is this something? It is not fourths, for six fourths are $1\frac{1}{2}$, and we had only 1 apple to begin with; neither is it eighths, for six eighths are $\frac{3}{4}$, which is less than 1 apple. By reducing the quarters to eighths, we have $\frac{2}{4} = \frac{4}{8}$, and adding the other 4 eighths, $4 + 4 = 8$ eighths. The result is correct, since $\frac{8}{8} = 1$. Or we can, in this case, reduce the eighths to quarters. Thus, $\frac{4}{8} = \frac{2}{4}$; whence, adding, $2 + 2 = 4$ quarters, a correct result, since $\frac{4}{4} = 1$.

Before adding, fractions should be reduced to a common denominator, preferably the *least* common denominator.

98. EXAMPLE.—Find the sum of $\frac{1}{2}$, $\frac{1}{4}$, and $\frac{5}{8}$.

SOLUTION.—The *least common denominator* or the *least number* which will contain all the *denominators* is 8.

$$\frac{1}{2} = \frac{4}{8}, \frac{1}{4} = \frac{2}{8}, \text{ and } \frac{5}{8} = \frac{5}{8}.$$

$$\frac{4}{8} + \frac{2}{8} + \frac{5}{8} = \frac{4+2+5}{8} = \frac{11}{8} = 1\frac{3}{8}. \text{ Ans.}$$

EXPLANATION.—As the *denominator* tells or indicates the names of the *parts*, the *numerators* only are added to obtain the total number of *parts* indicated by the *denominator*.

99. EXAMPLE.—What is the sum of $12\frac{5}{8}$, $14\frac{3}{4}$, and $7\frac{7}{16}$?

SOLUTION.—The least common denominator in this case is 16.

$$12\frac{5}{8} = 12\frac{10}{16}$$
$$14\frac{3}{4} = 14\frac{12}{16}$$
$$7\frac{7}{16} = 7\frac{7}{16}$$
$$\text{sum } 33 + \frac{29}{16} = 33 + 1\frac{13}{16} = 34\frac{13}{16}. \text{ Ans.}$$

The sum of the fractions $= \frac{29}{16}$ or $1\frac{13}{16}$, which added to the sum of the whole numbers $= 34\frac{13}{16}$.

EXAMPLE.—What is the sum of 17, $13\frac{9}{16}$, $\frac{7}{32}$, and $3\frac{1}{4}$?

SOLUTION.—The least common denominator is 32. $13\frac{9}{16} = 13\frac{18}{32}$, $3\frac{1}{4} = 3\frac{8}{32}$.

$$17$$
$$13\frac{18}{32}$$
$$\frac{7}{32}$$
$$3\frac{8}{32}$$
$$sum\ 33\frac{33}{32}.\ \ Ans.$$

100. Rule 9.—(*a*) *Reduce the given fractions to fractions having the least common denominator, and write the sum of the numerators over the common denominator.*

(*b*) *When there are mixed numbers and whole numbers, add the fractions first, and if their sum is an improper fraction, reduce it to a mixed number and add the whole number with the other whole numbers.*

EXAMPLES FOR PRACTICE.

101. Find the sum of

(*a*) $\frac{1}{3}$, $\frac{7}{24}$, $\frac{5}{8}$.
(*b*) $\frac{2}{3}$, $\frac{4}{15}$, $\frac{11}{45}$.
(*c*) $\frac{1}{2}$, $\frac{5}{8}$, $\frac{7}{16}$.
(*d*) $\frac{5}{8}$, $\frac{1}{4}$, $\frac{1}{8}$.
(*e*) $\frac{10}{17}$, $\frac{4}{85}$, $\frac{3}{5}$.
(*f*) $\frac{11}{12}$, $\frac{1}{6}$, $\frac{3}{4}$.
(*g*) $\frac{4}{11}$, $\frac{7}{22}$, $\frac{11}{12}$.
(*h*) $\frac{1}{4}$, $\frac{1}{8}$, $\frac{5}{8}$.

Ans. $\begin{cases} (a)\ 1\frac{7}{12}. \\ (b)\ 1\frac{4}{15}. \\ (c)\ 1\frac{9}{16}. \\ (d)\ 1\frac{1}{4}. \\ (e)\ 1\frac{22}{85}. \\ (f)\ 1\frac{5}{6}. \\ (g)\ 1\frac{7}{22}. \\ (h)\ 1. \end{cases}$

1. The weights of a number of castings were $412\frac{3}{4}$ lb., $270\frac{1}{2}$ lb., $1,020$ lb., $75\frac{1}{4}$ lb., and $68\frac{1}{2}$ lb. What was their total weight? *Ans.* 1,847 lb.

2. Four bolts are required, $2\frac{3}{4}$, $1\frac{7}{8}$, $2\frac{7}{16}$, and $1\frac{1}{2}$ inches long. How long a piece of iron will be required to cut them from, allowing $\frac{3}{4}$ of an inch altogether for cutting off and finishing the ends? *Ans.* $9\frac{5}{16}$ in.

SUBTRACTION OF FRACTIONS.

102. Fractions can not be *subtracted* without first reducing them to a *common denominator*. This can be shown in the same manner as in the case of addition of fractions.

EXAMPLE.—Subtract $\frac{3}{8}$ from $\frac{13}{16}$.

SOLUTION.—The common denominator is 16.

$$\frac{3}{8} = \frac{6}{16}.\ \ \frac{13}{16} - \frac{6}{16} = \frac{13-6}{16} = \frac{7}{16}.\ \ Ans.$$

ARITHMETIC. 31

103. Example.—From 7 take $\frac{3}{8}$.

Solution.—$1 = \frac{8}{8}$; therefore, since $7 = 6 + 1$ we see that $7 = 6 + \frac{8}{8}$, so that $6\frac{8}{8} - \frac{3}{8} = 6\frac{5}{8}$. Ans.

104. Example.—What is the difference between $17\frac{9}{16}$ and $9\frac{11}{32}$?

Solution.—The common denominator of the fractions is 32. $17\frac{9}{16} = 17\frac{18}{32}$.

$$\begin{array}{rr} minuend & 17\frac{18}{32} \\ subtrahend & 9\frac{11}{32} \\ \hline difference & 8\frac{7}{32} \end{array}$$ Ans.

105. Example.—From $9\frac{1}{4}$ take $4\frac{7}{16}$.

Solution.—The common denominator of the fractions is 16. $9\frac{1}{4} = 9\frac{4}{16}$.

$$\begin{array}{rcc} minuend & 9\frac{4}{16} & 8\frac{20}{16} \\ subtrahend & 4\frac{7}{16} \text{ or } & 4\frac{7}{16} \\ \hline difference & 4\frac{13}{16} & 4\frac{13}{16} \end{array}$$ Ans.

Explanation.—As the *fraction* in the *subtrahend* is greater than the fraction in the *minuend*, it *can not* be subtracted; therefore, *borrow* 1, or $\frac{16}{16}$, from the 9 in the *minuend* and *add* it to the $\frac{4}{16}$; $\frac{4}{16} + \frac{16}{16} = \frac{20}{16}$. $\frac{7}{16}$ from $\frac{20}{16} = \frac{13}{16}$. Since 1 was *borrowed* from 9, 8 *remains*; 4 from $8 = 4$; $4 + \frac{13}{16} = 4\frac{13}{16}$.

106. Example.—From 9 take $8\frac{3}{16}$.

$$\text{Solution.}—\begin{array}{rcc} minuend & 9 & 8\frac{16}{16} \\ subtrahend & 8\frac{3}{16} \text{ or } & 8\frac{3}{16} \\ \hline difference & \frac{13}{16} & \frac{13}{16} \end{array}$$ Ans.

Explanation.—As there is no *fraction* in the *minuend* from which to take the *fraction* in the *subtrahend*, borrow 1, or $\frac{16}{16}$, from 9. $\frac{3}{16}$ from $\frac{16}{16} = \frac{13}{16}$. Since 1 was *borrowed* from 9, only 8 is left. 8 from $8 = 0$.

107. Rule 10.—(*a*) *Reduce the fractions to fractions having a common denominator. Subtract one numerator from the other and place the remainder over the common denominator.*

(*b*) *When there are mixed numbers, subtract the fractions and whole numbers separately, and place the remainders side by side.*

(*c*) *When the fraction in the subtrahend is greater than the fraction in the minuend, borrow 1 from the whole number in the minuend and add it to the fraction in the minuend, from which subtract the fraction in the subtrahend.*

(*d*) *When the minuend is a whole number, borrow 1; reduce it to a fraction whose denominator is the same as the denominator of the fraction in the subtrahend, and place it over that fraction for subtraction.*

EXAMPLES FOR PRACTICE.

108. Subtract

(*a*) $\frac{1}{2}$ from $\frac{1}{4}$.
(*b*) $\frac{1}{7}$ from $\frac{1}{4}$.
(*c*) $\frac{3}{16}$ from $\frac{7}{16}$.
(*d*) $\frac{11}{16}$ from $\frac{15}{16}$.
(*e*) $\frac{1}{8}$ from $\frac{1}{4}$.
(*f*) $13\frac{1}{4}$ from $30\frac{1}{4}$.
(*g*) $12\frac{1}{4}$ from 27.
(*h*) $5\frac{1}{4}$ from 30.

Ans. $\begin{cases} (a)\ \frac{1}{4}. \\ (b)\ \frac{1}{14}. \\ (c)\ \frac{11}{16}. \\ (d)\ \frac{1}{16}. \\ (e)\ \frac{1}{8}. \\ (f)\ 17\frac{1}{4}. \\ (g)\ 14\frac{1}{4}. \\ (h)\ 24\frac{1}{4}. \end{cases}$

1. An engineer found that he had on hand $48\frac{1}{2}$ gallons of cylinder oil. During the following week he used $\frac{3}{4}$ of a gallon a day for three days, $\frac{7}{8}$ of a gallon on the fourth day, $1\frac{1}{8}$ of a gallon on the fifth day, and $\frac{1}{2}$ of a gallon on the sixth day. How much oil remained at the end of the week? Ans. $43\frac{11}{16}$ gallons.

2. The main line shaft of a manufacturing plant is run by an engine and water wheel. A test of the plant showed that the engine was capable of developing $251\frac{1}{4}$ H. P. (horsepower), and the water wheel, under full gate, $67\frac{1}{2}$ H. P. It was also found that the machinery consumed $210\frac{11}{16}$ H. P., and the friction of the shafting and belting was $32\frac{1}{4}$ H. P. How much power remained unused? Ans. $76\frac{11}{16}$ H. P.

MULTIPLICATION OF FRACTIONS.

109. In *multiplication* of fractions it is not necessary to *reduce* the *fractions* to fractions having a *common denominator*.

110. Multiplying the *numerator* or *dividing* the *denominator multiplies* the fraction.

ARITHMETIC.

EXAMPLE.—Multiply $\tfrac{3}{4} \times 4$.

SOLUTION.—

$$\tfrac{3}{4} \times 4 = \tfrac{3 \times 4}{4} = \tfrac{12}{4} = 3, \text{ Ans.}$$

or, $\quad \tfrac{3}{4} \times 4 = \tfrac{3}{4 \div 4} = \tfrac{3}{1} = 3.$ Ans.

The word "of" in multiplication of fractions means the same as $\times$, or times. Thus,

$$\tfrac{3}{4} \text{ of } 4 = \tfrac{3}{4} \times 4 = 3.$$

$$\tfrac{1}{8} \text{ of } \tfrac{5}{16} = \tfrac{1}{8} \times \tfrac{5}{16} = \tfrac{5}{128}.$$

EXAMPLE.—Multiply $\tfrac{3}{8}$ by 2.

SOLUTION.— $\quad \tfrac{3}{8} \times 2 = \tfrac{3 \times 2}{8} = \tfrac{6}{8} = \tfrac{3}{4},$ Ans.

or, $\quad \tfrac{3}{8} \times 2 = \tfrac{3}{8 \div 2} = \tfrac{3}{4}.$ Ans.

111. EXAMPLE.—What is the product of $\tfrac{4}{16}$ and $\tfrac{7}{8}$?

SOLUTION.— $\quad \tfrac{4}{16} \times \tfrac{7}{8} = \tfrac{4 \times 7}{16 \times 8} = \tfrac{28}{128} = \tfrac{7}{32},$

or, by cancelation, $\quad \dfrac{\cancel{4} \times 7}{\cancel{16}_{4} \times 8} = \tfrac{7}{4 \times 8} = \tfrac{7}{32}.$ Ans.

112. EXAMPLE.—What is $\tfrac{3}{8}$ of $\tfrac{4}{4}$ of $\tfrac{16}{32}$?

SOLUTION.— $\quad \dfrac{\cancel{4} \times 3 \times \cancel{16}}{8 \times \cancel{4} \times \cancel{32}_{2}} = \dfrac{3}{8 \times 2} = \tfrac{3}{16}.$ Ans.

113. EXAMPLE.—What is the product of $9\tfrac{3}{4}$ and $5\tfrac{5}{8}$?

SOLUTION.— $\quad 9\tfrac{3}{4} = \tfrac{39}{4};\ 5\tfrac{5}{8} = \tfrac{45}{8}.$

$\tfrac{39}{4} \times \tfrac{45}{8} = \dfrac{39 \times 45}{4 \times 8} = \tfrac{1755}{32} = 54\tfrac{27}{32}.$ Ans.

114. EXAMPLE.—Multiply $15\tfrac{1}{4}$ by 3.

SOLUTION.— $\quad\begin{array}{cc} 15\tfrac{1}{4} & 15\tfrac{1}{4} \\ \underline{\ \ 3\ \ } & \text{or } \underline{\ \ 3\ \ } \\ 47\tfrac{3}{4} & 45 + \tfrac{3}{4} \end{array} = 45 + 2\tfrac{3}{4} = 47\tfrac{3}{4}.$ Ans.

115. Rule 11.—(*a*) *Divide the product of the numerators by the product of the denominators. All factors common to the numerators and denominators should first be cast out by cancelation.*

ARITHMETIC.

(b) *To multiply one mixed number by another, reduce them both to improper fractions.*

(c) *To multiply a mixed number by a whole number, first multiply the fractional part by the multiplier, and if the product is an improper fraction, reduce it to a mixed number and add the whole number part to the product of the multiplier and whole number.*

EXAMPLES FOR PRACTICE.

116. Find the product of

(a) $7 \times \frac{3}{15}$.
(b) $14 \times \frac{5}{16}$.
(c) $\frac{21}{22} \times \frac{5}{14}$.
(d) $\frac{18}{27} \times 4$.
(e) $\frac{13}{18} \times 7$.
(f) $17\frac{11}{14} \times 7$.
(g) $\frac{101}{217} \times 32$.
(h) $\frac{15}{18} \times 14$.

Ans. $\begin{cases} (a) \ 1\frac{2}{15}. \\ (b) \ 4\frac{3}{8}. \\ (c) \ 1\frac{1}{4}. \\ (d) \ 2\frac{10}{27}. \\ (e) \ 7\frac{7}{18}. \\ (f) \ 125. \\ (g) \ 15. \\ (h) \ 7\frac{1}{2}. \end{cases}$

1. A single belt can transmit $107\frac{3}{4}$ H. P., but as it is desired to use more power, a double belt of the same width is substituted for it. Supposing the double belt to be capable of transmitting $\frac{1}{5}^{0}$ as much power as the single belt, how many H. P. can be used after the change? Ans. $153\frac{11}{14}$ H. P.

2. What is the weight of $2\frac{3}{5}$ miles of copper wire weighing $5\frac{3}{4}$ pounds per 100 feet? There are 5,280 feet in a mile.

Ans. $796\frac{13}{25}$ pounds.

3. The grate of a steam boiler contains $20\frac{1}{4}$ square feet. If the boiler burns $8\frac{3}{10}$ pounds of coal an hour per square foot of grate area, and can evaporate $7\frac{1}{2}$ pounds of water an hour per pound of coal burned, how many pounds of water are evaporated by the boiler in 1 hour? Ans. $1,276\frac{1}{8}$ pounds.

DIVISION OF FRACTIONS.

117. In *division* of fractions it is not necessary to *reduce* the *fractions* to fractions having a *common denominator*.

118. Dividing the *numerator* or *multiplying* the *denominator divides* the fraction.

ARITHMETIC.

EXAMPLE.—Divide $\frac{6}{8}$ by 3.
SOLUTION.—When *dividing* the *numerator*, we have

$$\frac{6}{8} \div 3 = \frac{6 \div 3}{8} = \frac{2}{8} = \frac{1}{4}. \quad \text{Ans.}$$

When *multiplying* the *denominator*, we have

$$\frac{6}{8} \div 3 = \frac{6}{8} \times 3 = \frac{6}{24} = \frac{1}{4}. \quad \text{Ans.}$$

EXAMPLE.—Divide $\frac{3}{16}$ by 2.
SOLUTION.— $\quad \frac{3}{16} \div 2 = \frac{3}{16 \times 2} = \frac{3}{32}. \quad$ Ans.

EXAMPLE.—Divide $\frac{14}{32}$ by 7.
SOLUTION.— $\quad \frac{14}{32} \div 7 = \frac{14 \div 7}{32} = \frac{2}{32} = \frac{1}{16}. \quad$ Ans.

119. To **invert** a fraction is to *turn it upside down;* that is, make the *numerator* and *denominator change places.* Invert $\frac{3}{4}$ and it becomes $\frac{4}{3}$.

120. EXAMPLE.—Divide $\frac{9}{16}$ by $\frac{3}{16}$.
SOLUTION.—1. The fraction $\frac{3}{16}$ is contained in $\frac{9}{16}$, 3 times, for the *denominators* are the same, and one *numerator* is *contained* in the other 3 times. 2. If we now *invert* the *divisor*, $\frac{3}{16}$, and *multiply*, the solution is

$$\frac{9}{16} \times \frac{16}{3} = \frac{\overset{3}{\cancel{9}} \times \cancel{16}}{\cancel{16} \times \cancel{3}} = 3. \quad \text{Ans.}$$

This gives the *same quotient* as in the first case.

121. EXAMPLE.—Divide $\frac{3}{8}$ by $\frac{1}{4}$.
SOLUTION.—We can not divide $\frac{3}{8}$ by $\frac{1}{4}$, as in the first case above, for the *denominators* are *not* the same; therefore, we must solve as in the second case.

$$\frac{3}{8} \div \frac{1}{4} = \frac{3}{8} \times \frac{4}{1} = \frac{3 \times \cancel{4}}{\underset{2}{\cancel{8}} \times 1} = \frac{3}{2} \text{ or } 1\frac{1}{2}. \quad \text{Ans.}$$

122. EXAMPLE.—Divide 5 by $\frac{10}{16}$.
SOLUTION.—$\frac{10}{16}$ inverted becomes $\frac{16}{10}$,

$$5 \times \frac{16}{10} = \frac{\cancel{5} \times \overset{8}{\cancel{16}}}{\underset{\cancel{2}}{\cancel{10}}} = 8. \quad \text{Ans.}$$

ARITHMETIC.

123. EXAMPLE.—How many times is $3\frac{3}{4}$ contained in $7\frac{7}{16}$?

SOLUTION.— $3\frac{3}{4} = \frac{15}{4}$; $7\frac{7}{16} = \frac{119}{16}$.

$\frac{15}{4}$ inverted becomes $\frac{4}{15}$.

$$\frac{119}{16} \times \frac{4}{15} = \frac{119 \times \cancel{4}}{\cancel{16} \times 15} = \frac{119}{60} = 1\frac{59}{60}.$$ Ans.

124. Rule 12.—*Invert the divisor and proceed as in multiplication.*

125. We have learned that a line placed between two numbers indicates that the number above the line is to be divided by the number below it. Thus, $\frac{18}{3}$ shows that 18 is to be divided by 3. This is also true if a fraction or a fractional expression be placed above or below a line.

$\frac{9}{\frac{3}{8}}$ means that 9 is to be divided by $\frac{3}{8}$; $\dfrac{3 \times 7}{\dfrac{8+4}{16}}$ means that 3×7 is to be divided by the value of $\dfrac{8+14}{16}$.

$\dfrac{\frac{1}{4}}{\frac{3}{8}}$ is the same as $\frac{1}{4} \div \frac{3}{8}$.

It will be noticed that there is a heavy line between the 9 and the $\frac{3}{8}$. This is necessary, since otherwise there would be nothing to show whether 9 is to be divided by $\frac{3}{8}$, or $\frac{3}{8}$ by 8. Whenever a heavy line is used, as in the above case, it indicates that *all above the line* is to be divided by *all below it*.

EXAMPLES FOR PRACTICE.

126. Divide

(a) 15 by $6\frac{2}{3}$.
(b) 30 by $\frac{3}{4}$.
(c) 172 by $\frac{4}{5}$.
(d) $\frac{11}{16}$ by $1\frac{7}{16}$.
(e) $\frac{19}{27}$ by $14\frac{3}{8}$.
(f) $\frac{141}{27}$ by $17\frac{1}{4}$.
(g) $\frac{11}{8}$ by $1\frac{18}{5}$.
(h) $\frac{121}{16}$ by $72\frac{1}{3}$.

Ans.
(a) $2\frac{1}{4}$.
(b) 40.
(c) 215.
(d) $\frac{11}{16}$.
(e) $1\frac{13}{18}$.
(f) $\frac{71}{297}$.
(g) $\frac{55}{144}$.
(h) $\frac{44}{381}$.

1. A $\frac{3}{8}$-inch boiler plate containing 24 square feet of surface weighs $362\frac{4}{10}$ pounds. What is its weight per square foot? Ans. $15\frac{1}{10}$ pounds.

ARITHMETIC.

2. A certain boiler has $927\frac{1}{2}$ square feet of heating surface, which is equal to 35 times the area of the grate. What is the area of the grate in square feet? Ans. $26\frac{1}{2}$ square feet.

3. If the distance around the rim of a locomotive drive wheel is $13\frac{1}{12}$ feet, how many revolutions will the wheel make in traveling 682 feet? Ans. $52\frac{20}{157}$ revolutions.

127. Whenever an expression like one of the three following ones is obtained, it may always be simplified by transposing the denominator from *above* to *below* the line, or from *below* to *above*, as the case may be, taking care, however, to indicate that the denominator when so transferred is a multiplier.

1. $\dfrac{\frac{3}{4}}{9} = \dfrac{3}{9 \times 4} = \dfrac{3}{36} = \dfrac{1}{12}$; for, regarding the fraction above the heavy line as the numerator of a fraction whose denominator is 9, $\dfrac{\frac{3}{4} \times 4}{9 \times 4} = \dfrac{3}{9 \times 4}$, as before.

2. $\dfrac{9}{\frac{3}{4}} = \dfrac{9 \times 4}{3} = 12$. The proof is the same as in the first case.

3. $\dfrac{\frac{5}{3}}{\frac{3}{4}} = \dfrac{5 \times 4}{3 \times 9} = \dfrac{20}{27}$; for, regarding $\frac{5}{3}$ as the numerator of a fraction whose denominator is $\frac{3}{4}$, $\dfrac{\frac{5}{3} \times 9}{\frac{3}{4} \times 9} = \dfrac{5}{\frac{3 \times 9}{4}}$; and $\dfrac{\frac{5}{3 \times 9} \times 4}{\frac{3 \times 9}{4} \times 4} = \dfrac{5 \times 4}{3 \times 9} = \dfrac{20}{27}$, as above.

This principle may be used to great advantage in cases like $\dfrac{\frac{1}{4} \times 310 \times \frac{27}{1} \times 72}{40 \times 4\frac{1}{2} \times 5\frac{1}{6}}$. Reducing the mixed numbers to fractions, the expression becomes $\dfrac{\frac{1}{4} \times 310 \times \frac{27}{1} \times 72}{40 \times \frac{9}{2} \times \frac{31}{6}}$. Now transferring the denominators of the fractions and canceling,

$$\dfrac{1 \times 310 \times 27 \times 72 \times 2 \times 6}{40 \times 9 \times 31 \times 4 \times 12} = \dfrac{1 \times \cancel{310}^{10} \times \cancel{27}^{3} \times \cancel{72}^{6,3} \times \cancel{2} \times \cancel{6}^{3}}{\cancel{40}^{4,2} \times \cancel{9} \times \cancel{31} \times \cancel{4} \times \cancel{12}^{2}} =$$

$\dfrac{27}{2} = 13\frac{1}{2}$.

ARITHMETIC.

Greater exactness in results can usually be obtained by using this principle than can be obtained by reducing the fractions to decimals. The principle, however, should not be employed *if a sign of addition or subtraction occurs either above or below the dividing line.*

DECIMALS.

128. Decimals are *tenth* fractions; that is, the parts of a unit are expressed on the scale of ten, as *tenths, hundredths, thousandths*, etc.

129. The *denominator*, which is always ten or a multiple of ten, as 10, 100, 1,000, etc., is *not* expressed as it would be in common fractions, by writing it under the *numerator*, with a line between them; as, $\frac{3}{10}$, $\frac{3}{100}$, $\frac{3}{1000}$. The denominator is always understood, the numerator consisting of the figures on the right of the *unit* figure. In order to distinguish the unit figure, a period (.), called the **decimal point,** is placed between the unit figure and the next figure on the right. The decimal point may be regarded in two ways: first, as indicating that the number on the right is the numerator of a fraction whose denominator is 10, 100, 1,000, etc.; and, second, as a part of the Arabic system of notation, each figure on the right being 10 times as large as the next succeeding figure, and 10 times as small as the next preceding figure, serving merely to point out the unit figure.

130. The *reading* of a *decimal number* depends upon the *number* of *decimal places* in it, or the *number* of *figures* to the *right* of the unit figure.

The first figure to the right of the unit figure expresses *tenths.*

The second figure to the right of the unit figure expresses *hundredths.*

The third figure to the right of the unit figure expresses *thousandths.*

The fourth figure to the right of the unit figure expresses *ten-thousandths.*

The fifth figure to the right of the unit figure expresses *hundred-thousandths.*

ARITHMETIC.

The sixth figure to the right of the unit figure expresses *millionths*.

Thus:

.3	=	$\frac{3}{10}$	=	3 tenths.
.03	=	$\frac{3}{100}$	=	3 hundredths.
.003	=	$\frac{3}{1000}$	=	3 thousandths.
.0003	=	$\frac{3}{10000}$	=	3 ten-thousandths.
.00003	=	$\frac{3}{100000}$	=	3 hundred-thousandths.
.000003	=	$\frac{3}{1000000}$	=	3 millionths.

The first figure to the right of the unit figure is called the *first* **decimal place**; the second figure, the *second decimal place*, etc. We see in the above that the *number* of *decimal places* in a decimal equals the *number of ciphers* to the *right* of the figure 1 in the *denominator* of its *equivalent fraction*. This fact kept in mind will be of much assistance in reading and writing decimals.

Whatever may be written to the *left* of a *decimal point* is a whole number. The decimal point affects only the figures to its *right*.

When a *whole number* and *decimal* are written together, the expression is a *mixed number*. Thus, 8.12 and 17.25 are mixed numbers.

The relation of decimals and whole numbers to each other is clearly shown by the following table:

hundreds of millions.	tens of millions.	millions.	hundreds of thousands.	tens of thousands.	thousands.	hundreds.	tens.	units.	decimal point.	tenths.	hundredths.	thousandths.	ten-thousandths.	hundred-thousandths.	millionths.	ten-millionths.	hundred-millionths.
9	8	7	6	5	4	3	2	1	.	2	3	4	5	6	7	8	9

The figures to the *left* of the *decimal point* represent *whole numbers;* those to the *right* are *decimals*.

In *both* the decimals and whole numbers, the *units* place is made the *starting point* of notation and numeration. The *decimals decrease* on the scale of *ten* to the *right*, and the *whole* numbers *increase* on the scale of *ten* to the *left*. The *first* figure to the *left* of units is *tens*, and the *first* figure to the *right* of units is *tenths*. The *second* figure to the *left* of units is *hundreds*, and the *second* figure to the *right* is *hundredths*. The *third* figure to the *left* is *thousands*, and the *third* to the *right* is *thousandths*, and so on ; the *whole* numbers on the *left* and the *decimals* on the *right*. The figures equally distant from units place correspond in name. The *decimals* have the ending *ths*, which distinguishes them from *whole* numbers. The following is the numeration of the number in the above table : Nine hundred eighty-seven million, six hundred fifty-four thousand, three hundred twenty-one, and twenty-three million, four hundred fifty-six thousand, seven hundred eighty-nine hundred millionths.

The *decimals* increase to the *left* on a scale of *ten*, the same as *whole* numbers, for if you begin at 4-*thousandths*, in the table, you see that the next figure is *hundredths*, which is ten times as great, and the next *tenths*, or ten times the *hundredths*, and so on through both decimals and whole numbers.

131. *Annexing* or *taking away* a *cipher* at the *right* of a *decimal* does *not* affect its value.

.5 is $\frac{5}{10}$; .50 is $\frac{50}{100}$, but $\frac{5}{10} = \frac{50}{100}$; therefore, .5 = .50.

132. *Inserting* a *cipher* between a *decimal* and the *decimal point divides* the decimal by 10.

.5 = $\frac{5}{10}$; $\frac{5}{10} \div 10 = \frac{5}{100}$ = .05.

133. *Taking away* a *cipher* from the *left* of a *decimal multiplies* the *decimal* by 10.

.05 = $\frac{5}{100}$; $\frac{5}{100} \times 10 = \frac{5}{10}$ = .5.

ADDITION OF DECIMALS.

134. The only respect in which *addition* of *decimals* differs from *addition* of *whole numbers* is in the placing of the numbers to be added.

Whole numbers begin at units and increase on the scale of 10 to the left. Decimals decrease on the scale of 10 to the right. Whole numbers are to the left of the decimal point and decimals are to the right of it. In whole numbers the *right hand* side of a column of figures to be added must be in line, and in decimals the *left hand* side must be in line, which brings the decimal points directly under each other.

whole numbers	*decimals*	*mixed numbers*
342	.342	342.032
4234	.4234	4234.5
26	.26	26.6782
3	.03	3.06
sum 4605 Ans.	*sum* 1.0554 Ans.	*sum* 4606.2702 Ans.

135. EXAMPLE.—What is the sum of 242, .36, 118.725, 1.005, 6, and 100.1?

SOLUTION.—
```
        242.
          .36
        118.725
          1.005
          6.
        100.1
 sum    468.190  Ans.
```

136. Rule 13.—*Place the numbers to be added so that the decimal points will be directly under each other. Add as in whole numbers, and place the decimal point in the sum directly under the decimal points above.*

EXAMPLES FOR PRACTICE.

137. Find the sum of

(*a*) .2143, .105, 2.3042, and 1.1417.
(*b*) 783.5, 21.473, .2101, and .7816.
(*c*) 21.781, 138.72, 41.8738, .72, and 1.413.
(*d*) .3724, 104.15, 21.417, and 100.042.
(*e*) 200.172, 14.105, 12.1465, .705, and 7.2.
(*f*) 1,427.16, .244, .32, .032, and 10.0041.
(*g*) 2,473.1, 41.65, .7243, 104.067, and 21.073.
(*h*) 4,107.2, .00375, 21.716, 410.072, and .0345.

Ans.
(*a*) 3.7652.
(*b*) 805.9647.
(*c*) 204.5078.
(*d*) 225.9814.
(*e*) 234.3285.
(*f*) 1,437.7601.
(*g*) 2,640.6143.
(*h*) 4,539.02625.

1. The estimated weights of the parts of a return tubular boiler were as follows: Shell, 3,626 lb.; tubes, 3,564.5 lb.; manhole cover, ring, and yoke, 270.34 lb.; stays, etc., 1,089.4 lb.; steam nozzles,

236.07 lb.; handhole covers and yokes, 120.25 lb.; feed pipe, 34.75 lb.; boiler supports, 350.6 lb. What was the total estimated weight of the boiler? Ans. 9,291.91 lb.

2. A bill for engine room supplies had the following items: 1 waste can, $8.30; 20 feet of 4-inch belting, $11.20; 1 pipe wrench, $1.65; 12 pounds of waste, $0.84; 5 gallons of cylinder oil, $8.75; 20 gallons of machine oil, $24. How much did the bill amount to? Ans. $54.74.

SUBTRACTION OF DECIMALS.

138. For the same reason as in addition of decimals, the *left hand* figures of *decimal numbers* are placed in line and the *decimal points* under each other.

EXAMPLE.—Subtract .132 from .3063.
SOLUTION.—
  minuend .3063
  subtrahend .132
  difference .1743 Ans.

139. EXAMPLE.—What is the difference between 7.895 and .725?
SOLUTION.—
  minuend 7.895
  subtrahend .725
  difference 7.170 or 7.17 Ans.

140. EXAMPLE.—Subtract .625 from 11.
SOLUTION.—
  minuend 11.000
  subtrahend .625
  difference 10.375

141. Rule 14.—*Place the subtrahend under the minuend, so that the decimal points will be directly under each other. Subtract as in whole numbers, and place the decimal point in the remainder, directly under the decimal points above.*

When the figures in the decimal part of the subtrahend extend beyond those in the minuend, place ciphers in the minuend above them and subtract as before.

EXAMPLES FOR PRACTICE.

142. From
(*a*) 407.385 take 235.0004.
(*b*) 22.718 take 1.7042.
(*c*) 1,368.17 take 13.6817.
(*d*) 70.00017 take 7.000017.

Ans. { (*a*) 172.3846.
      (*b*) 21.0138.
      (*c*) 1,354.4883.
      (*d*) 63.000153.

ARITHMETIC. 43

(e) 630.630 take .6304.
(f) 421.73 take 217.162.
(g) 1.000014 take .00001.
(h) .783652 take .542314.

Ans.
{
(e) 629.9996.
(f) 204.568.
(g) 1.000004.
(h) .241338.
}

1. If the temperature of steam at 5 pounds pressure is 227.964 degrees, and at 100 pounds pressure 337.874 degrees, how many degrees warmer is the steam at the higher pressure?   Ans. 109.91 degrees.

2. The outside diameter of 2¼-inch wrought-iron pipe is 2.87 inches and the inside diameter 2.46 inches. How thick is the pipe?
Ans. .41 ÷ 2 = .205 inch.

3. In a cistern that will hold 326.5 barrels of water there are 178.625 barrels. How much does it lack of being full?   Ans. 147.875 barrels.

4. A wrought-iron rod is 2.53 inches in diameter. What must be the thickness of metal turned off so that the rod will be 2.495 inches in diameter?   Ans. .035 ÷ 2 = .0175.

MULTIPLICATION OF DECIMALS.

143. In multiplication of decimals, we do not place the *decimal points* directly *under* each other as in addition and subtraction. We pay *no* attention for the time being to the decimal points. Place the multiplier under the multiplicand, so that the *right-hand* figure of the one is under the *right-hand* figure of the other, and proceed exactly as in multiplication of whole numbers. After multiplying, *count the number of decimal places in both multiplicand and multiplier, and point off the same number in the product.*

EXAMPLE.—Multiply .825 by 13.

SOLUTION.—
```
   multiplicand    .825
     multiplier     13
                  -----
                   2475
                    825
                  -----
       product   10.725  Ans.
```

In this example there are three decimal places in the multiplicand and none in the multiplier; therefore, 3 decimal places are pointed off in the product.

144. EXAMPLE.—What is the product of 426 and the decimal .005?

SOLUTION.—
   multiplicand 4 2 6
   multiplier   .0 0 5
   product 2.1 3 0 or 2.13 Ans.

In this example there are 3 decimal places in the multiplier and none in the multiplicand; therefore, 3 decimal places are pointed off in the product.

145. It is *not* necessary to multiply by the ciphers on the *left* of a *decimal;* they merely determine the number of decimal places. Ciphers to the *right* of a decimal should be removed, as they only make more figures to deal with, and do not change the value.

146. EXAMPLE.—Multiply 1.205 by 1.15.

SOLUTION.—
   multiplicand 1.2 0 5
   multiplier   1.1 5
         6 0 2 5
        1 2 0 5
       1 2 0 5
   product 1.3 8 5 7 5 Ans.

In this example there are 3 decimal places in the multiplicand, and 2 in the multiplier, therefore $3 + 2$, or 5, decimal places must be pointed off in the product.

147. EXAMPLE.—Multiply .232 by .001.

SOLUTION.—
   multiplicand .2 3 2
   multiplier   .0 0 1
   product .0 0 0 2 3 2 Ans.

In this example we multiply the multiplicand by the digit in the multiplier, which makes 232 in the product, but since there are 3 decimal places in each, the multiplier and the multiplicand, we must prefix 3 ciphers to the 232, to make $3 + 3$, or 6, decimal places in the product.

148. Rule 15.—*Place the multiplier under the multiplicand, disregarding the position of the decimal points. Multiply as in whole numbers, and in the product point off as*

ARITHMETIC.

many decimal places as there are decimal places in both multiplier and multiplicand, prefixing ciphers if necessary.

EXAMPLES FOR PRACTICE.

149. Find the product of

(*a*) .000492 × 4.1418.
(*b*) 4,003.2 × 1.2.
(*c*) 78.6531 × 1.03.
(*d*) .3685 × .042.
(*e*) 178,352 × .01.
(*f*) .00045 × .0045.
(*g*) .714 × .00002.
(*h*) .00004 × .008.

Ans.
(*a*) .0020377656.
(*b*) 4,803.84.
(*c*) 81.012693.
(*d*) .015477.
(*e*) 1,783.52.
(*f*) .000002025.
(*g*) .00001428.
(*h*) .00000032.

1. The stroke of an engine was found by measurement to be 2.987 feet. How many feet will the cross-head pass over in 600 revolutions?
Ans. 3,584.4 feet.

2. If a steam pump delivers 2.39 gallons of water per stroke and runs at 51 strokes a minute, how many gallons of water would it pump in 58½ minutes? Ans. 7,130.565 gallons.

3. Wishing to obtain the weight of a connecting-rod from a drawing, it was calculated that the rod contained 294.8 cubic inches of wrought iron, 63.5 cubic inches of brass, and 10.4 cubic inches of babbitt. Assuming the weight of wrought iron to be .278 pound per cubic inch, of brass .303 pound, and of babbitt .264 pound, what was the weight of the rod? Ans. 103.94 pounds.

DIVISION OF DECIMALS.

150. In division of decimals we pay *no* attention to the *decimal point* until *after* the *division* is performed. The *number of decimal places* in the *dividend must equal* (*be made to equal by annexing ciphers*) the *number of decimal places in the divisor*. Divide *exactly* as in whole numbers. Subtract the *number* of *decimal places* in the *divisor* from the *number* of *decimal places* in the *dividend*, and *point off* as many *decimal* *places in the quotient as there are units in the remainder thus found.*

EXAMPLE.—Divide .625 by 25.

SOLUTION.—
$$\begin{array}{r}\text{divisor dividend quotient}\\25\,)\,.6\,2\,5\,(\,.0\,2\,5 \text{ Ans.}\\5\,0\\\hline1\,2\,5\\1\,2\,5\\\hline\text{remainder }0\end{array}$$

In this example there are no decimal places in the divisor, and 3 decimal places in the dividend; therefore, there are 3 minus 0, or 3, decimal places in the quotient. One cipher has to be prefixed to the 25, to make the 3 decimal places.

151. EXAMPLE.—Divide 6.035 by .05.

SOLUTION.—
$$\begin{array}{r}\text{divisor dividend quotient}\\.0\,5\,)\,6.0\,3\,5\,(\,1\,2\,0.7 \text{ Ans.}\\5\\\hline1\,0\\1\,0\\\hline3\,5\\3\,5\\\hline\text{remainder }0\end{array}$$

In this example we divide by 5, as if the cipher were not before it. There is one more decimal place in the dividend than in the divisor; therefore, one decimal place is pointed off in the quotient.

152. EXAMPLE.—Divide .125 by .005.

SOLUTION.—
$$\begin{array}{r}\text{divisor dividend quotient}\\.0\,0\,5\,)\,.1\,2\,5\,(\,2\,5 \text{ Ans.}\\1\,0\\\hline2\,5\\2\,5\\\hline\text{remainder }0\end{array}$$

In this example there are the same number of decimal places in the dividend as in the divisor; therefore, the quotient has no decimal places, and is a whole number.

ARITHMETIC. 47

153. EXAMPLE.—Divide 326 by .25.

SOLUTION.—
```
        divisor  dividend  quotient
          .25 ) 3 2 6.0 0 ( 1 3 0 4   Ans.
                2 5
                ———
                  7 6
                  7 5
                  ———
                    1 0 0
                    1 0 0
                    ———
         remainder     0
```

In this problem two ciphers were annexed to the dividend, to make the number of decimal places equal to the number in the divisor. The quotient is a whole number.

154. EXAMPLE.—Divide .0025 by 1.25.

SOLUTION.—
```
          1.25 ) .0 0 2 5 0 ( .0 0 2   Ans.
                  2 5 0
                  ———
       remainder     0
```

EXPLANATION.—In this example we are to divide .0025 by 1.25. Consider the dividend as a whole number, or 25 (disregarding the two ciphers at its left, for the present); also, consider the divisor as a whole number, or 125. It is clearly evident that the dividend 25 will not contain the divisor 125; we must, therefore, annex one cipher to the 25, thus making the dividend 250. 125 is contained twice in 250, so we place the figure 2 in the quotient. In pointing off the decimal places in the quotient, it must be remembered that there were only four decimal places in the dividend; but one cipher was annexed, thereby making $4+1$, or 5, decimal places. Since there are 5 decimal places in the dividend and 2 decimal places in the divisor, we must point off $5-2$, or 3, decimal places in the quotient. In order to point off 3 decimal places, two ciphers must be prefixed to the figure 2, thereby making .002 the quotient. It is not necessary to consider the ciphers at the left of a decimal when dividing, except when determining the position of the decimal point in the quotient.

155. Rule 16.—(**a**) *Place the divisor to the left of the dividend and proceed as in division of whole numbers, and in the* *quotient, point off as many decimal places as the number of decimal places in the dividend exceed those in the divisor, prefixing ciphers to the quotient, if necessary.*

(b) *If in dividing one number by another there be a remainder, the remainder can be placed over the divisor, as a fractional part of the quotient, but it is generally better to annex ciphers to the remainder, and continue dividing until there are 3 or 4 decimal places in the quotient, and then if there still be a remainder, terminate the quotient by the plus sign* (+), *which shows that it can be carried further.*

156. EXAMPLE.—What is the quotient of 199 divided by 15?

SOLUTION.—

$$15 \overline{)199} (13 + \tfrac{4}{15} \quad \text{Ans.}$$
$$\underline{15}$$
$$49$$
$$\underline{45}$$
$$\text{remainder } \ 4$$

Or, $\quad 15 \overline{)199.000} (13.266 + \quad$ **Ans.**
$$\underline{15}$$
$$49$$
$$\underline{45}$$
$$40$$
$$\underline{30}$$
$$100$$
$$\underline{90}$$
$$100$$
$$\underline{90}$$
$$\text{remainder } \ 10$$

$$13 \tfrac{4}{15} = 13.266 +$$
$$\tfrac{4}{15} = .266 +$$

It very frequently happens as in the above example, that the division will never terminate. In such cases, decide to how many decimal places the division is to be carried, and carry the work one place further. If the last figure of the quotient thus obtained is 5 or a greater number, increase the preceding figure by 1, and write after it the minus sign (−), thus indicating that the quotient is not quite as large as indicated; if the figure thus obtained is less than 5, write the plus sign (+) after the quotient, thus indicating that the number is slightly greater than as indicated. In the last example, had it been desired to obtain the answer correct to four decimal places, the work would have been carried to five places, obtaining 13.26666, and the answer would have been given as 13.2667—. This remark applies to any other calculation involving decimals, when it is desired to omit some of the figures in the decimal. Thus, if it is desired to retain three decimal places in the number .2471253, it would be expressed as .247 +; if it was desired to retain five decimal places, it would be expressed as .24713 —. Both the + and — signs are frequently omitted; they are seldom used outside of arithmetic, except in exact calculations, when it is desired to call particular attention to the fact that the result obtained is not quite exact.

EXAMPLES FOR PRACTICE.

157. Divide
(a) 101.6688 by 2.36.
(b) 187.12264 by 123.107.
(c) .08 by .008.
(d) .0003 by 3.75.
(e) .0144 by .024.
(f) .00375 by 1.25.
(g) .004 by 400.
(h) .4 by .008.

Ans. $\begin{cases} (a)\ 43.08. \\ (b)\ 1.52. \\ (c)\ 10. \\ (d)\ .00008. \\ (e)\ .6. \\ (f)\ .003. \\ (g)\ .00001. \\ (h)\ 50. \end{cases}$

1. In a steam engine test of an hour's duration, the horsepower developed was found to be as follows, at 10-minute intervals: 25.73, 25.64, 26.13, 25.08, 24.20, 26.7, 26.34. What was the average horsepower?   Ans. 25.6886, average.

NOTE.—Add the different horsepowers together and divide by the number of tests, or 7.

2. There are 31.5 gallons in a barrel. How many barrels are there in 2,787.75 gallons?   Ans. 88.5 barrels.

3. A car load of 18.75 tons of coal cost $60.75. How much was it worth per ton?   Ans. $3.24 per ton.

4. A keg of $\frac{7}{16}$ by 1⅜-inch boiler rivets weighs 100 pounds and contains 595 rivets. What is the weight of one of the rivets?
  Ans. .168 pound.

ARITHMETIC.

TO REDUCE A FRACTION TO A DECIMAL.

158. EXAMPLE.—$\frac{3}{4}$ equals what decimal?
SOLUTION.—
$$4)\overline{3.00}$$
$$.75$$
, or $\frac{3}{4} = .75$. Ans.

EXAMPLE.—What decimal is equal to $\frac{7}{8}$?
SOLUTION.—
$$8)7.000(.875$$
$$\underline{64}$$
$$60$$
$$\underline{56}\quad \text{or } \frac{7}{8} = .875.\text{ Ans.}$$
$$40$$
$$\underline{40}$$
$$0$$

159. Rule 17.—*Annex ciphers to the numerator and divide by the denominator. Point off as many decimal places in the quotient as there are ciphers annexed.*

EXAMPLES FOR PRACTICE.

160. Reduce the following common fractions to decimals:

(a) $\frac{15}{32}$.
(b) $\frac{7}{8}$.
(c) $\frac{21}{32}$.
(d) $\frac{51}{64}$.   Ans.
(e) $\frac{4}{25}$.
(f) $\frac{5}{8}$.
(g) $\frac{10}{200}$.
(h) $\frac{4}{1000}$.

$\begin{cases} (a)\ .46875. \\ (b)\ .875. \\ (c)\ .65625. \\ (d)\ .796875. \\ (e)\ .16. \\ (f)\ .625. \\ (g)\ .05. \\ (h)\ .004. \end{cases}$

161. To reduce inches to decimal parts of a foot:

EXAMPLE.—What decimal part of a foot is 9 inches?

SOLUTION.—Since there are 12 inches in one foot, 1 inch is $\frac{1}{12}$ of a foot, and 9 inches is $9 \times \frac{1}{12}$ or $\frac{9}{12}$ of a foot. This reduced to a decimal by the above rule, shows what decimal part of a foot 9 inches is.

$$12)9.00(.75 \text{ of a foot. Ans.}$$
$$\underline{84}$$
$$60$$
$$\underline{60}$$
$$0$$

ARITHMETIC.

162. **Rule 18.**—(*a*) *To reduce inches to decimal parts of a foot, divide the number of inches by 12.*

(*b*) *Should the resulting decimal be an unending one and it is desired to terminate the division at some point, say the fourth decimal place, carry the division one place farther, and if the fifth figure is 5 or greater, increase the fourth figure by one and omit the sign* +.

EXAMPLES FOR PRACTICE.

163. Reduce to the decimal part of a foot:

(*a*) 3 in.
(*b*) 4¼ in.
(*c*) 5 in.   Ans. $\begin{cases} (a)\ .25. \\ (b)\ .375. \\ (c)\ .4167. \\ (d)\ .5521. \\ (e)\ .9167. \end{cases}$
(*d*) 6⅝ in.
(*e*) 11 in.

1. The lengths of belting required to connect three countershafts with the main line shaft were found with a tape measure to be 27 ft. 4 in., 23 ft. 8 in., and 38 ft. 6 in. How many feet of belting were necessary? Ans. 89.5 ft.

2. The stroke of an engine is 14 inches. What is the length of the crank measured from the center of shaft to center of crank-pin in feet? Ans. .5833+ foot.

3. A steam pipe fitted with an expansion joint was found to expand 1.668 inches when steam was admitted to it. How much was its expansion in decimal parts of a foot? Ans. .139 foot.

TO REDUCE A DECIMAL TO A FRACTION.

164. EXAMPLE.—Reduce .125 to a fraction.

SOLUTION.—$.125 = \frac{125}{1000} = \frac{5}{40} = \frac{1}{8}$. Ans.

EXAMPLE.—Reduce .875 to a fraction.

SOLUTION.—$.875 = \frac{875}{1000} = \frac{35}{40} = \frac{7}{8}$. Ans.

165. **Rule 19.**—*Under the figures of the decimal, place the digit 1 with as many ciphers at its right as there are decimal places in the decimal, and reduce the resulting fraction to its lowest terms by dividing both numerator and denominator by the same number.*

EXAMPLES FOR PRACTICE.

166. Reduce the following to common fractions:

(a) .125.
(b) .625.
(c) .3125.
(d) .04.
(e) .06.
(f) .75.
(g) .15625.
(h) .875.

Ans. $\begin{cases} (a)\ \frac{1}{8}. \\ (b)\ \frac{5}{8}. \\ (c)\ \frac{5}{16}. \\ (d)\ \frac{1}{25}. \\ (e)\ \frac{3}{50}. \\ (f)\ \frac{3}{4}. \\ (g)\ \frac{5}{32}. \\ (h)\ \frac{7}{8}. \end{cases}$

TO EXPRESS A DECIMAL APPROXIMATELY AS A FRACTION HAVING A GIVEN DENOMINATOR.

167. EXAMPLE.—Express .5827 in 64ths.

SOLUTION.—$.5827 \times \frac{64}{64} = \frac{37.2928}{64}$, say $\frac{37}{64}$.

Hence, $.5827 = \frac{37}{64}$, nearly. Ans.

EXAMPLE.—Express .3917 in 12ths.

SOLUTION.—$.3917 \times \frac{12}{12} = \frac{4.7004}{12}$, say $\frac{5}{12}$.

Hence, $.3917 = \frac{5}{12}$, nearly. Ans.

168. **Rule 20.**—*Reduce 1 to a fraction having the given denominator. Multiply the given decimal by the fraction so obtained, and the result will be the fraction required.*

EXAMPLES FOR PRACTICE.

169. Express

(a) .625 in 8ths.
(b) .3125 in 16ths.
(c) .15625 in 32ds.
(d) .77 in 64ths.
(e) .81 in 48ths.
(f) .923 in 96ths.

Ans. $\begin{cases} (a)\ \frac{5}{8}. \\ (b)\ \frac{5}{16}. \\ (c)\ \frac{5}{32}. \\ (d)\ \frac{49}{64}. \\ (e)\ \frac{39}{48}. \\ (f)\ \frac{89}{96}. \end{cases}$

ARITHMETIC. 53

170. The sign for dollars is $. It is read dollars. $25 is read 25 dollars.

Since there are 100 cents in a dollar, one cent is 1-one-hundredth of a dollar; the first two figures of a decimal part of a dollar represent *cents*. Since a mill is $\frac{1}{10}$ of a cent, or $\frac{1}{1000}$ of a dollar, the third figure represents mills.

Thus, $25.16 is read twenty-five dollars and sixteen cents; $25.168 is read twenty-five dollars, sixteen cents and eight mills.

171. The **vinculum** ——, **parenthesis** ( ), **bracket** [ ], and **brace** { } are called **symbols of aggregation**, and are used to include numbers which are to be considered together; thus, $13 \times \overline{8-3}$, or $13 \times (8-3)$, shows that 3 is to be taken from 8 before multiplying by 13.

$$13 \times (8-3) = 13 \times 5 = 65. \quad \text{Ans.}$$
$$13 \times \overline{8-3} = 13 \times 5 = 65. \quad \text{Ans.}$$

When the vinculum or parenthesis is not used, we have
$$13 \times 8 - 3 = 104 - 3 = 101. \quad \text{Ans.}$$

172. In any series of numbers connected by the signs +, −, ×, and ÷, the operations indicated by the signs must be performed in order from left to right, *except* that no addition or subtraction may be performed if a sign of multiplication or division *follows* the number on the *right* of a sign of addition or subtraction, until the indicated multiplication or division has been performed. In all cases the sign of multiplication takes the precedence, the reason being that when two or more numbers or expressions are connected by the sign of multiplication, the numbers thus connected are regarded as factors of the product indicated, and not as separate numbers.

EXAMPLE.—What is the value of $4 \times 24 - 8 + 17$?

SOLUTION.—Performing the operations in order from left to right, $4 \times 24 = 96$; $96 - 8 = 88$; $88 + 17 = 105$. Ans.

173. EXAMPLE.—What is the value of the following expression: $1,296 \div 12 + 160 - 22 \times 3\frac{1}{2} = ?$

SOLUTION.—$1,296 \div 12 = 108$; $108 + 160 = 268$; here we cannot subtract 22 from 268 because the sign of multiplication *follows* 22; hence, multiplying 22 by $3\frac{1}{2}$, we get 77, and $268 - 77 = 191$. Ans.

Had the above expression been written $1,296 \div 12 + 160 - 22 \times 3\frac{1}{2} \div 7 + 25$, it would have been necessary to have divided $22 \times 3\frac{1}{2}$ by 7 before subtracting, and the final result would have been $22 \times 3\frac{1}{2} = 77$; $77 \div 7 = 11$; $268 - 11 = 257$; $257 + 25 = 282$. Ans. In other words, it is necessary to perform *all* of the multiplication or division included between the signs $+$ and $-$, or $-$ and $+$, before adding or subtracting. Also, had the expression been written $1,296 \div 12 + 160 - 24\frac{1}{2} \div 7 \times 3\frac{1}{2} + 25$, it would have been necessary to have multiplied $3\frac{1}{2}$ by 7 before dividing $24\frac{1}{2}$, since the sign of multiplication takes the precedence, and the final result would have been $3\frac{1}{2} \times 7 = 24\frac{1}{2}$; $24\frac{1}{2} \div 24\frac{1}{2} = 1$; $268 - 1 = 267$; $267 + 25 = 292$. Ans.

It likewise follows that if a succession of multiplication and division signs occurs, the indicated operations must not be performed in order, from left to right—the multiplication must be performed first. Thus, $24 \times 3 \div 4 \times 2 \div 9 \times 5 = \frac{1}{3}$. Ans. In order to obtain the same result that would be obtained by performing the indicated operations in order, from left to right, symbols of aggregation must be used. Thus, by using two vinculums, the last expression becomes $24 \times \overline{3 \div 4} \times \overline{2 \div 9} \times 5 = 20$, the same result that would be obtained by performing the indicated operations in order, from left to right.

EXAMPLES FOR PRACTICE.

174. Find the values of the following expressions:

(*a*) $(8 + 5 - 1) + 4$.
(*b*) $5 \times 24 - 32$.
(*c*) $5 \times 24 + 15$.
(*d*) $144 - 5 \times 24$.
(*e*) $(1,691 - 540 + 559) + 3 \times 57$.
(*f*) $2,080 + 120 - 80 \times 4 - 1,670$.
(*g*) $\overline{(90 + 60 + 25)} \times 5 - 29$.
(*h*) $\overline{90 + 60} + 25 \times 5$.

Ans. $\begin{cases} (a) & 3. \\ (b) & 88. \\ (c) & 8. \\ (d) & 24. \\ (e) & 10. \\ (f) & 210. \\ (g) & 1. \\ (h) & 1.2. \end{cases}$

ARITHMETIC.
(CONTINUED.)

PERCENTAGE.

175. Percentage is the process of calculating by *hundredths*.

176. The *term* **per cent.** is an abbreviation of the Latin words *per centum*, which mean *by the hundred*. A certain per cent. of a number is the number of hundreds of that number which is indicated by the number of units in the per cent. Thus, 6 per cent. of 125 is $125 \times \frac{6}{100} = 7.5$; 25 per cent. of 80 is $80 \times \frac{25}{100} = 20$; 43 per cent. of 432 pounds is $432 \times \frac{43}{100} = 185.76$ pounds.

177. The **sign** of per cent. is %, and is read *per cent*. Thus, 6% is read *six per cent.*; 12½% is read *twelve and one-half per cent.*, etc.

When expressing the per cent. of a number to use in calculations it is customary to express it decimally instead of fractionally. Thus, instead of expressing 6%, 25%, and 43% as $\frac{6}{100}$, $\frac{25}{100}$, and $\frac{43}{100}$, it is usual to express them as .06, .25, and .43.

The following table will show how any per cent. can be expressed either as a decimal or as a fraction:

TABLE 3.

Per Cent.	Decimal.	Fraction.	Per Cent.	Decimal.	Fraction.
1%	.01	$\frac{1}{100}$	150 %	1.50	$\frac{150}{100}$ or $1\frac{1}{2}$
2%	.02	$\frac{2}{100}$ or $\frac{1}{50}$	500 %	5.00	$\frac{500}{100}$ or 5
5%	.05	$\frac{5}{100}$ or $\frac{1}{20}$	¼%	.0025	$\frac{\frac{1}{4}}{100}$ or $\frac{1}{400}$
10%	.10	$\frac{10}{100}$ or $\frac{1}{10}$	½%	.005	$\frac{\frac{1}{2}}{100}$ or $\frac{1}{200}$
25%	.25	$\frac{25}{100}$ or $\frac{1}{4}$	1½%	.015	$\frac{1\frac{1}{2}}{100}$ or $\frac{3}{200}$
50%	.50	$\frac{50}{100}$ or $\frac{1}{2}$	8⅓%	.08⅓	$\frac{8\frac{1}{3}}{100}$ or $\frac{1}{12}$
75%	.75	$\frac{75}{100}$ or $\frac{3}{4}$	12½%	.125	$\frac{12\frac{1}{2}}{100}$ or $\frac{1}{8}$
100%	1.00	$\frac{100}{100}$ or 1	16⅔%	.16⅔	$\frac{16\frac{2}{3}}{100}$ or $\frac{1}{6}$
125%	1.25	$\frac{125}{100}$ or $1\frac{1}{4}$	62½%	.625	$\frac{62\frac{1}{2}}{100}$ or $\frac{5}{8}$

178. The names of the different elements used in percentage are: the *base*, the *rate per cent.*, the *percentage*, the *amount*, and the *difference*.

179. The **base** is the number on which the per cent. is computed.

180. The **rate** is the number of hundredths of the base to be taken.

181. The **percentage** is the part, or number of *hundredths*, of the base indicated by the rate; or the percentage is the result obtained by multiplying the base by the rate.

Thus, when it is stated that 7% of $25 is $1.75, $25 is the base, 7% is the rate, and $1.75 is the percentage.

182. The **amount** is the sum of the base and percentage.

183. The **difference** is the remainder obtained by subtracting the percentage from the base.

Thus, if a man has $180, and he earns 6% more, he will have altogether $180 + $180 × .06, or $180 + $10.80 = $190.80. Here $180 is the base, 6% the rate, $10.80 the percentage, and $190.80 the *amount*.

Again, if an engine of 125 horsepower uses 16% of it in overcoming friction and other resistances, the amount left for performing useful work is 125 − 125 × .16 = 125 − 20 = 105 horsepower. Here 125 is the base, 16% the rate, 20 the percentage, and 105 the *difference*.

184. From the foregoing it is evident that, to find the percentage, the base must be multiplied by the rate. Hence the following

Rule 21.—*To find the percentage, multiply the base by the rate, expressed decimally.*

EXAMPLE.—Out of a lot of 300 boiler tubes 76% were used in a boiler. How many tubes were used?

SOLUTION.—76%, the rate, expressed decimally, is .76; the base is 300; hence, the number of tubes used, or the percentage, is by the above rule   300 × .76 = 228 tubes. Ans.

Expressing the rule as a formula, we have

$$percentage = base \times rate.$$

ARITHMETIC.

185. When the percentage and rate are given, the base may be found by dividing the percentage by the rate. For, suppose that 12 is 6%, or $\frac{6}{100}$, of some number; then, 1%, or $\frac{1}{100}$, of the number is $12 \div 6$ or 2. Consequently, if $2 = 1\%$, or $\frac{1}{100}$, 100%, or $\frac{100}{100} = 2 \times 100 = 200$. But, since the same result may be arrived at by dividing 12 by .06, for $12 \div .06 = 200$ it follows that

Rule 22.—*When the percentage and rate are given, to find the base, divide the percentage by the rate, expressed decimally.*

Formula, *base = percentage ÷ rate.*

EXAMPLE.—76% of a lot of boiler tubes are used in the construction of a boiler. If the number of tubes used were 228, how many tubes were in the lot?

SOLUTION.—Here 228 is the percentage, and 76%, or .76, is the rate; hence, applying the rule,

$228 \div .76 = 300$ tubes. Ans.

186. When the base and percentage are given to find the rate, the rate may be found, expressed decimally, by dividing the percentage by the base. For suppose that it is desired to find what per cent. 12 is of 200. 1% of 200 is $200 \times .01 = 2$. Now, if 1% is 2, 12 is evidently as many per cent. as the number of times that 2 is contained in 12, or $12 \div 2 = 6\%$. But the same result may be obtained by dividing 12, the percentage, by 200, the base, since $12 \div 200 = .06 = 6\%$. Hence,

Rule 23.—*When the percentage and base are given, to find the rate, divide the percentage by the base, and the result will be the rate, expressed decimally.*

Formula, *rate = percentage ÷ base.*

EXAMPLE.—Out of a lot of 300 boiler tubes, 228 were used. What per cent. of the total number was used?

SOLUTION.—Here 300 is the base and 228 the percentage; hence, applying rule,

rate $= 228 \div 300 = .76 = 76\%$. Ans.

EXAMPLE.—What per cent. of 875 is 25?

SOLUTION.—Here 875 is the base, and 25 is the percentage; hence, applying rule,

$25 \div 875 = .02\frac{4}{7} = 2\frac{4}{7}\%$. Ans.

PROOF.—$875 \times .02\frac{4}{7} = 25$.

ARITHMETIC.

EXAMPLES FOR PRACTICE.

187. What per cent. of
(a) 360 is 90?
(b) 900 is 360?
(c) 125 is 25?
(d) 150 is 750?
(e) 280 is 112?
(f) 400 is 200?
(g) 47 is 94?
(h) 500 is 250?

Ans. $\begin{cases} (a)\ 25\% \\ (b)\ 40\% \\ (c)\ 20\% \\ (d)\ 500\% \\ (e)\ 40\% \\ (f)\ 50\% \\ (g)\ 200\% \\ (h)\ 50\% \end{cases}$

188. The amount may be found when the base and rate are given, by multiplying the base by 1 plus the rate, expressed decimally. For suppose that it is desired to find the amount when 200 is the base and 6% is the rate. The percentage is $200 \times .06 = 12$, and, according to definition, Art. **182,** the amount is $200 + 12 = 212$. But the same result may be obtained by multiplying 200 by $1 + .06$, or 1.06, since $200 \times 1.06 = 212$. Hence,

Rule 24.—*When the base and rate are given, to find the amount, multiply the base by 1 plus the rate, expressed decimally.*

Formula, *amount* $= base \times (1 + rate)$.

EXAMPLE.—If a man earned $725 in a year, and the next year 10% more, how much did he earn the second year?

SOLUTION.—Here 725 is the base and 10% is the rate, and the amount is required. Hence, applying the rule,

$725 \times 1.10 = \$797.50$. Ans.

189. When the base and rate are given, the difference may be found by multiplying the base by 1 minus the rate, expressed decimally. For suppose that it is desired to find the difference when the base is 200 and the rate is 6%. The percentage is $200 \times .06 = 12$; and, according to definition, Art. **183,** the difference $= 200 - 12 = 188$. But the same result may be obtained by multiplying 200 by $1 - .06$, or .94, since $200 \times .94 = 188$. Hence,

Rule 25.—*When the base and rate are given, to find the difference, multiply the base by 1 minus the rate, expressed decimally.*

Formula, *difference* $= base \times (1 - rate)$.

ARITHMETIC. 59

EXAMPLE.—Out of a lot of 300 boiler tubes all but 24% were used in one boiler; how many tubes were used?

SOLUTION.—Here 300 is the base, 24% is the rate, and it is desired to find the difference. Hence, applying the rule.

$$300 \times (1 - .24) = 228 \text{ tubes. Ans.}$$

190. When the amount and rate are given, the base may be found by dividing the amount by 1 plus the rate. For suppose that it is known that 212 equals some number increased by 6% of itself. Then, it is evident that 212 equals 106% of the number (base) that it is desired to find. Consequently, if $212 = 106\%$, $1\% = \frac{212}{106} = 2$, and $100\% = 2 \times 100 = 200 = $ the base. But the same result may be obtained by dividing 212 by $1 + .06$ or 1.06, since $212 \div 1.06 = 200$. Hence,

Rule 26.—*When the amount and rate are given, to find the base, divide the amount by 1 plus the rate, expressed decimally.*

Formula, $base = amount \div (1 + rate)$.

EXAMPLE.—The theoretical discharge of a certain pump when running at a piston speed of 100 feet per minute is 278,910 gallons per day of 10 hours. Owing to leakage and other defects this value is 25% greater than the actual discharge. What is the actual discharge?

SOLUTION.—Here 278,910 equals the actual discharge (base) increased by 25% of itself. Consequently, 278,910 is the amount, 25% is the rate, and applying rule,

actual discharge $= 278,910 \div 1.25 = 223,128$ gallons. Ans.

191. When the difference and rate are given, the base may be found by dividing the difference by 1 minus the rate. For suppose that 188 equals some number less 6% of itself. Then, 188 evidently equals $100 - 6 = 94\%$ of some number. Consequently, if $188 = 94\%$, $1\% = 188 \div 94 = 2$, and $100\% = 2 \times 100 = 200$. But the same result may be obtained by dividing 188 by $1 - .06$, or .94, since $188 \div .94 = 200$. Hence,

Rule 27.—*When the difference and rate are given, to find the base, divide the difference by 1 minus the rate, expressed decimally.*

Formula, $base = difference \div (1 - rate)$.

ARITHMETIC.

EXAMPLE.—From a lot of boiler tubes 76% were used in the construction of a boiler. If there were 72 tubes unused, how many tubes were in the lot?

SOLUTION.—Here 72 is the difference and 76% is the rate. Applying rule,

$$72 \div (1 - .76) = 300 \text{ tubes. Ans.}$$

EXAMPLE.—The theoretical number of foot-pounds of work per minute required to operate a boiler feed-pump is 127,344. If 30% of the total number actually required be allowed for friction, leakage, etc., how many foot-pounds are actually required to work the pump?

SOLUTION.—Here the number actually required is the base; hence, 127,344 is the difference, and 30% is the rate. Applying the rule,

$$127,344 \div (1 - .30) = 181,920 \text{ foot-pounds. Ans.}$$

192. EXAMPLE.—A certain chimney gave a draft of 2.76 inches of water. By increasing the height 20 feet the draft was increased to 3 inches of water. What was the gain per cent.?

SOLUTION.—Here it is evident that 3 inches is the amount and that 2.76 inches is the base. Consequently, $3 - 2.76 = .24$ inch is the percentage, and it is required to find the rate. Hence, applying rule **23**,

$$\text{gain per cent.} = .24 \div 2.76 = .087 = 8.7\%. \text{ Ans.}$$

193. EXAMPLE.—A certain chimney gave a draft of 3 inches of water. After an economizer had been put in the draft was reduced to 1.2 inches of water. What was the loss per cent.?

SOLUTION.—Here it is evident 1.2 inches is the difference (since it equals 3 inches diminished by a certain per cent. loss of itself) and 3 inches is the base. Consequently, $3 - 1.2 = 1.8$ inches is the percentage. Hence, applying rule **23**,

$$\text{loss per cent.} = 1.8 \div 3 = .60 = 60\%. \text{ Ans.}$$

194. To find the gain or loss per cent.:

Rule 28.—*Find the difference between the initial and final values; divide this difference by the initial value.*

EXAMPLE.—If a man buys a steam engine for $1,860 and some time afterwards purchases a condenser for 25% of the cost of the engine, does he gain or lose, and how much per cent., if he sells both engine and condenser for $2,100?

SOLUTION.—The cost of the condenser was $1,860 × .25 = $465; consequently, the total initial value, or cost, was $1,860 + $465 = $2,325. Since he sold them for $2,100, he lost $2,325 − $2,100 = $225. Hence, applying rule,

$$225 \div 2,325 = .0968 = 9.68\% \text{ loss. Ans.}$$

ARITHMETIC.

EXAMPLES FOR PRACTICE.

195. Solve the following:

(a) What is 12½% of $900?   Ans. (a) $112.50.
(b)   "   " ⅞% " 627?   (b) 5.016.
(c)   "   " 33⅓% " 54?   (c) 18.
(d) 101 is 68¼% of what number?   (d) 146¹⁹⁄₂₇.
(e) 784 " 83⅓% "   "   "   (e) 940.8.
(f) What % of 960 is 160?   (f) 16⅔%.
(g)   "   "   " $3,606 is $450¾? (g) 12½%.
(h)   "   "   " 280 is 112?   (h) 40%.

1. A steam plant consumed an average of 3,640 pounds of coal per day. The engineer made certain alterations which resulted in a saving of 250 pounds per day. What was the per cent. of coal saved?

Ans. 7%, nearly.

2. If the speed of an engine running at 126 revolutions per minute should be increased 6½%, how many revolutions per minute would it then make?   Ans. 134.19 revolutions.

3. The list price of an engine was $1,400; of a boiler, $1,150, and of the necessary fittings for the two, $340. If 25% discount was allowed on the engine, 22% on the boiler, and 12½% on the fittings, what was the actual cost of the plant?   Ans. $2,244.50.

4. If I lend a man $1,100, and this is 18½% of the amount that I have on interest, how much money have I on interest?   Ans. $5,945.95.

5. A test showed that an engine developed 190.4 horsepower, 15% of which was consumed in friction. How much power was available for use?   Ans. 161.84 H. P.

6. By adding a condenser to a steam engine, the power was increased 14%, and the consumption of coal per horsepower per hour was decreased 20%. If the engine could originally develop 50 horsepower and required 3¼ pounds of coal per horsepower per hour, what would be the total weight of coal used in an hour, with the condenser, assuming the engine to run full power?   Ans. 159.6 pounds.

DENOMINATE NUMBERS.

196. A denominate number is a concrete number, and may be either simple or compound, as 8 quarts, 5 feet, ten inches, etc.

197. A simple denominate number consists of units of but one denomination, as 16 cents, 10 hours, 5 dollars, etc.

198. A **compound denominate number** consists of units of two or more denominations of a similar kind, as 3 yards, 2 feet, 1 inch ; 3 pounds, 5 ounces.

199. In **whole numbers** and in **decimals**, the *law* of increase and decrease is on the scale of 10, but in **compound** or **denominate numbers** the scale varies.

MEASURES.

200. A **measure** is a *standard unit* established by *law* or *custom*, by which *quantity* of any kind is measured. The *standard unit* of **dry measure** is the Winchester bushel; of **weight**, the pound; of **liquid measure**, the gallon, etc.

201. Measures are of six kinds:

1. Extension.
2. Weight.
3. Capacity.
4. Time.
5. Angles.
6. Money or value.

MEASURES OF EXTENSION.

202. **Measures of extension** are used in measuring lengths, distances, surfaces, and solids.

LINEAR MEASURE.
TABLE 4.

	Abbreviation.	in.	ft.	yd.	rd.	fur.
12 inches (in.) = 1 foot	ft.	36 =	3			
3 feet = 1 yard	yd.	198 =	16½ =	5½		
5½ yards = 1 rod	rd.	7,920 =	660 =	220 =	40	
40 rods = 1 furlong	fur.	63,360 =	5,280 =	1,760 =	320 =	8
8 furlongs = 1 mile	mi.					

SQUARE MEASURE.
TABLE 5.

144 square inches (sq. in.)	= 1 square foot	sq. ft.
9 square feet	= 1 square yard	sq. yd.
30¼ square yards	= 1 square rod	sq. rd.
160 square rods	= 1 acre	A.
640 acres	= 1 square mile	sq. mi.

sq. mi.	A.	sq. rd.	sq. yd.	sq. ft.	sq. in.
1	= 640	= 102,400	= 3,097,600	= 27,878,400	= 4,014,489,600

ARITHMETIC. 63

CUBIC MEASURE.
TABLE 6.

1728 cubic inches (cu. in.) . . .	= 1 cubic foot	cu. ft.
27 cubic feet	= 1 cubic yard	cu. yd.
128 cubic feet	= 1 cord	cd.
24¾ cubic feet	= 1 perch	P.

cu. yd.   cu. ft.   cu. in.
1   =   27   =   46,656

MEASURES OF WEIGHT.

AVOIRDUPOIS WEIGHT.
TABLE 7.

16 ounces (oz.)	= 1 pound	lb.
100 pounds	= 1 hundred-weight . . .	cwt.
20 cwt., or 2,000 lb.	= 1 ton	T.

T.   cwt.   lb.   oz.
1 = 20 = 2,000 = 32,000

203. The ounce is divided into halves, quarters, etc. Avoirdupois weight is used for weighing coarse and heavy articles.

LONG TON TABLE.
TABLE 8.

16 ounces	= 1 pound	lb.
112 pounds	= 1 hundred-weight . . .	cwt.
20 cwt., or 2,240 lb.	= 1 ton	T.

In all the calculations throughout this and the succeeding volumes, 2,000 pounds will be considered one ton, unless the long ton (2,240 pounds) is especially mentioned.

TROY WEIGHT.
TABLE 9.

24 grains (gr.)	= 1 pennyweight	pwt.
20 pennyweight	= 1 ounce	oz.
12 ounces	= 1 pound	lb.

lb.   oz.   pwt.   gr.
1 = 12 = 240 = 5,760

204. Troy weight is used in weighing gold and silver ware, jewels, etc. It is used by jewelers.

MEASURES OF CAPACITY.

LIQUID MEASURE.
TABLE 10.

4 gills (gi.)	= 1 pint	pt.
2 pints	= 1 quart	qt.
4 quarts	= 1 gallon	gal.
31½ gallons	= 1 barrel	bbl.
63 gallons	= 1 hogshead	hhd.

hhd.	bbl.	gal.	qt.	pt.	gi.
1	= 2	= 63	= 252	= 504	= 2,016

DRY MEASURE.
TABLE 11.

2 pints (pt.)	= 1 quart	qt.
8 quarts	= 1 peck	pk.
4 pecks	= 1 bushel	bu.

bu.	pk.	qt.	pt.
1	= 4	= 32	= 64

MEASURE OF TIME.
TABLE 12.

60 seconds (sec.)	= 1 minute	min.
60 minutes	= 1 hour	hr.
24 hours	= 1 day	da.
7 days	= 1 week	wk.
365 days } 12 months }	= 1 common year	yr.
366 days	= 1 leap year	
100 years	= 1 century	

NOTE.—It is customary to consider one month as 30 days.

MEASURE OF ANGLES OR ARCS.
TABLE 13.

60 seconds (″)	= 1 minute	′
60 minutes	= 1 degree	°
90 degrees	= 1 right angle or quadrant	∟
360 degrees	= 1 circle	cir.

clr.			
1	= 360°	= 21,600′	= 1,296,000″

ARITHMETIC.

MEASURE OF MONEY.
UNITED STATES MONEY.
TABLE 14.

10 mills (m.) . . . . = 1 cent . . . . ct.
10 cents . . . . = 1 dime . . . . d.
10 dimes . . . . = 1 dollar . . . . $.
10 dollars . . . . = 1 eagle . . . . E.

E. $ d.   ct.   m.
1 = 10 = 100 = 1,000 = 10,000

MISCELLANEOUS TABLE.
TABLE 15.

12 things are 1 dozen.
12 dozen are 1 gross.
12 gross are 1 great gross.
2 things are 1 pair.
20 things are 1 score.
1 league is 3 miles.
1 fathom is 6 feet.

1 meter is nearly 39.37 inches.
1 hand is 4 inches.
1 palm is 3 inches.
1 span is 9 inches.
24 sheets are 1 quire.
20 quires, or 480 sheets, are 1 ream.
1 bushel contains 2,150.4 cubic in.

1 U. S. standard gallon (also called a wine gallon) contains 231 cubic in
1 U. S. standard gallon of water weighs 8.355 pounds, nearly.
1 cubic foot of water contains 7.481 U. S. standard gallons, nearly.
1 British imperial gallon weighs 10 pounds.

It will be of great advantage to the student to carefully memorize all of the above tables.

REDUCTION OF DENOMINATE NUMBERS.

205. Reduction of denominate numbers is the process of changing their denomination without changing their value. They may be changed from a higher to a lower denomination or from a lower to a higher—either is reduction. As,

2 hours = 120 minutes.
32 ounces = 2 pounds.

206. Principle.—Denominate numbers are changed to *lower* denominations by *multiplying*, and to *higher* denominations by *dividing*.

To reduce denominate numbers to lower denominations:

207. Example.—Reduce 5 yd. 2 ft. 7 in. to inches.

Solution.—

	yd.	ft.	in.
	5	2	7
	3		

15 ft.
2 ft.
―――
17 ft.
12
―――
34
17
―――
204 in.
7 in.
―――
211 inches. Ans.

Explanation.—Since there are 3 feet in 1 yard, in 5 yards there are 5 × 3, or 15 feet, and 15 feet plus 2 feet = 17 feet. There are 12 inches in a foot; therefore, 12 × 17 = 204 inches, and 204 inches plus 7 inches = 211 inches = number of inches in 5 yards 2 feet and 7 inches. Ans.

208. Example.—Reduce 6 hours to seconds.

Solution.—

6 hours.
60
―――
360 minutes.
60
―――
21600 seconds. Ans.

Explanation.—As there are 60 minutes in one hour, in six hours there are 6 × 60, or 360 minutes; as there are no minutes to add, we multiply 360 minutes by 60, to get the number of seconds.

209. In order to avoid mistakes, if any denomination be omitted, represent it by a cipher. Thus, before reducing 3 rods 6 inches to inches, insert a cipher for yards and a cipher for feet; as,

rd.	yd.	ft.	in.
3	0	0	6

210. Rule.—*Multiply the number representing the highest denomination by the number of units in the next lower*

ARITHMETIC.

required to make one of the higher denomination, and to the product add the number of given units of that lower denomination. Proceed in this manner until the number is reduced to the required denomination.

EXAMPLES FOR PRACTICE.

211. Reduce

    (*a*) 4 rd. 2 yd. 2 ft. to ft.
    (*b*) 4 bu. 3 pk. 2 qt. to qt.
    (*c*) 13 rd. 5 yd. 2 ft. to ft.
    (*d*) 5 mi. 100 rd. 10 ft. to ft.
    (*e*) 8 lb. 4 oz. 6 pwt. to gr.
    (*f*) 52 hhd. 24 gal. 1 pt. to pt.
    (*g*) 5 cir. 16° 20' to minutes.
    (*h*) 14 bu. to qt.

Ans.
    (*a*) 74 ft.
    (*b*) 154 qt.
    (*c*) 231.5 ft.
    (*d*) 28,060 ft.
    (*e*) 48,144 gr.
    (*f*) 26,401 pt.
    (*g*) 108,980'.
    (*h*) 448 qt.

To reduce lower to higher denominations:

212. EXAMPLE.—Reduce 211 in. to higher denominations.

SOLUTION.—
$$\begin{array}{r}12\,)\,211\text{ in.}\\ \hline 3\,)\,17\text{ ft.}+7\text{ in.}\\ \hline 5\text{ yd.}+2\text{ ft. Ans.}\end{array}$$

EXPLANATION.—There are 12 inches in 1 foot; therefore, 211 divided by 12 = 17 feet and 7 inches over. There are 3 ft. in 1 yd.; therefore, 17 ft. divided by 3 = 5 yd. and 2 ft. over. The last quotient and the two remainders constitute the answer, 5 yd. 2 ft. 7 in.

213. EXAMPLE.—Reduce 14,135 gi. to higher denominations.

SOLUTION.—
$$\begin{array}{r}4\,)\,14135\\ \hline 2\,)\,3533\text{ pt. 3 gi.}\\ \hline 4\,)\,1766\text{ qt. 1 pt.}\\ \hline 441\text{ gal. 2 qt.}\end{array}$$

$$\begin{array}{r}31.5\,)\,441.0\,(\,14\text{ bbl.}\\ 315\\ \hline 1260\\ 1260\end{array}$$

EXPLANATION.—There are 4 gi. in 1 pt., and in 14,135 gi. there are as many pints as 4 is contained in 14,135, or 3,533 pt. and 3 gi. remaining. There are 2 pt. in 1 qt., and in 3,533 pt. there are 1,766 qt. and 1 pt. remaining. There are 4 qt. in 1 gal., and in 1,766 qt. there are 441 gal. and 2 qt. remaining. There are 31½ gal. in 1 bbl., and in 441 gal. there are 14 bbl.

The last quotient and the three remainders constitute the answer, 14 bbl. 2 qt. 1 pt. 3 gi.

214. Rule 30.—*Divide the number representing the denomination given, by the number of units of this denomination required to make one unit of the next higher denomination. The remainder will be of the same denomination, but the quotient will be of the next higher. Divide this quotient by the number of units of its denomination required to make one unit of the next higher. Continue until the highest denomination is reached, or until there is not enough of a denomination left to make one of the next higher. The last quotient and the remainders constitute the required result.*

EXAMPLES FOR PRACTICE.

215. Reduce to units of higher denominations:
(*a*) 7,460 sq. in.; (*b*) 7,580 sq. yd.; (*c*) 148,760 cu. in.; (*d*) 17,651″; (*e*) 8,000 gi.; (*f*) 36,450 lb.

Ans.
(*a*) 5 sq. yd. 6 sq. ft. 116 sq. in.
(*b*) 1 A. 90 sq. rd. 17 sq. yd. 4 sq. ft. 72 sq. in.
(*c*) 3 cu. yd. 5 cu. ft. 152 cu. in.
(*d*) 4° 54′ 11″.
(*e*) 3 hhd. 61 gal.
(*f*) 18 T. 4 cwt. 50 lb.

ADDITION OF DENOMINATE NUMBERS.

216. Example.—Find the sum of 3 cwt. 46 lb. 12 oz.; 8 cwt. 12 lb. 13 oz.; 12 cwt. 50 lb. 13 oz.; 27 lb. 4 oz.

Solution.—

	T.	cwt.	lb.	oz.
	0	3	46	12
	0	8	12	13
	0	12	50	13
	0	0	27	4
	1	4	37	10 Ans.

EXPLANATION.—Begin to add at the right-hand column: $4 + 13 + 13 + 12 = 42$ ounces; as 16 ounces make 1 pound, 42 ounces $\div 16 = 2$ and a remainder of 10 ounces, or 2 pounds and 10 ounces. Place 10 ounces under ounce column and add 2 pounds to the next or pound column. Then, $2 + 27 + 50 + 12 + 46 = 137$ pounds; as 100 pounds make a hundred-weight, $137 \div 100 = 1$ hundred-weight and a remainder of 37 pounds. Place the 37 under the pound column, and add 1 hundred-weight to the next or hundred-weight column. Next, $1 + 12 + 8 + 3 = 24$ hundred-weight. 20 hundred-weight make a ton; therefore $24 \div 20 = 1$ ton and 4 hundred-weight remaining. Hence, the sum is 1 ton 4 hundred-weight 37 pounds 10 ounces. Ans.

217. EXAMPLE.—What is the sum of 2 rd. 3 yd. 2 ft. 5 in.; 6 rd. 1 ft. 10 in.; 17 rd. 11 in.; 4 yd. 1 ft.?

SOLUTION.—

	rd.	yd.	ft.	in.
	2	3	2	5
	6	0	1	10
	17	0	0	11
	0	4	1	0
	26	3½	0	2
or	26	3	1	8 Ans.

EXPLANATION.—The sum of the numbers in the first column $= 26$ inches, or 2 feet and 2 inches remaining. The sum of the numbers in the next column plus 2 feet $= 6$ feet, or 2 yards and 0 feet remaining. The sum of the next column plus 2 yards $= 9$ yards, or $9 \div 5\frac{1}{2} = 1$ rod and $3\frac{1}{2}$ yards remaining. The sum of the next column plus 1 rod $= 26$ rods. To avoid fractions in the sum, the $\frac{1}{2}$ yard is reduced to 1 foot and 6 inches, which added to 26 rods 3 yards 0 feet and 2 inches $= 26$ rods 3 yards 1 foot 8 inches. Ans.

218. EXAMPLE.—What is the sum of 47 ft. and 3 rd. 2 yd. 2 ft. 10 in.?

SOLUTION.—When 47 ft. is reduced it equals 2 rd. 4 yd. 2 ft., which can be added to 3 rd. 2 yd. 2 ft. 10 in. Thus,

	rd.	yd.	ft.	in.
	3	2	2	10
	2	4	2	0
	6	1½	1	10
or	6	2	0	4 Ans.

219. Rule 31.—*Place the numbers so that like denominations are under each other. Begin at the right-hand column, and add. Divide the sum by the number of units of this denomination required to make one unit of the next higher. Place the remainder under the column added, and carry the quotient to the next column. Continue in this manner until the highest denomination given is reached.*

EXAMPLES FOR PRACTICE.

220. What is the sum of

(*a*) 25 lb. 7 oz. 15 pwt. 23 gr.; 17 lb. 16 pwt.; 15 lb. 4 oz. 12 pwt.; 18 lb. 16 gr.; 10 lb. 2 oz. 11 pwt. 16 gr.?

(*b*) 9 mi. 13 rd. 4 yd. 2 ft.; 16 rd. 5 yd. 1 ft. 5 in.; 16 mi. 2 rd. 3 in.; 14 rd. 1 yd. 9 in.?

(*c*) 3 cwt. 46 lb. 12 oz.; 12 cwt. 9¼ lb.; 2¼ cwt. 21⅜ lb?

(*d*) 10 yr. 8 mo. 5 wk. 3 da.; 42 yr. 6 mo. 7 da.; 7 yr. 5 mo. 18 wk. 4 da.; 17 yr. 17 da.?

(*e*) 17 tons 11 cwt. 49 lb. 14 oz.; 16 tons 47 lb. 13 oz.; 20 tons 13 cwt. 14 lb. 6 oz.; 11 tons 4 cwt. 16 lb. 12 oz.?

(*f*) 14 sq. yd. 8 sq. ft. 19 sq. in.; 105 sq. yd. 16 sq. ft. 240 sq. in.; 42 sq. yd. 28 sq. ft. 165 sq. in.?

Ans.
(*a*) 86 lb. 3 oz. 16 pwt. 7 gr.
(*b*) 25 mi. 47 rd. 1 ft. 5 in.
(*c*) 18 cwt. 2 lb. 14 oz.
(*d*) 78 yr. 1 mo. 3 wk. 3 da.
(*e*) 65 tons 9 cwt. 28 lb. 13 oz.
(*f*) 167 sq. yd. 136 sq. in.

SUBTRACTION OF DENOMINATE NUMBERS.

221. EXAMPLE.—From 21 rd. 2 yd. 2 ft. 6¼ in., take 9 rd. 4 yd. 10¼ in.

SOLUTION.—

rd.	yd.	ft.	in.
21	2	2	6¼
9	4	0	10¼
11	3¼	1	8¼ Ans.

EXPLANATION.—Since $10\frac{1}{4}$ inches can not be taken from $6\frac{1}{4}$ inches, we must borrow 1 foot, or 12 inches, from the 2 feet in the next column and add it to the $6\frac{1}{2}$. $6\frac{1}{2} + 12 = 18\frac{1}{2}$. $18\frac{1}{2}$ inches $- 10\frac{1}{4}$ inches $= 8\frac{1}{4}$ inches. Then, 0 foot from the 1 remaining foot $= 1$ foot. 4 yards can not be taken from 2 yards; therefore, we borrow 1 rod, or $5\frac{1}{2}$ yards, from 21 rods and add it to 2. $2 + 5\frac{1}{2} = 7\frac{1}{2}$; $7\frac{1}{2} - 4 = 3\frac{1}{2}$ yards. 9 rods from 20 rods $= 11$ rods. Hence, the remainder is 11 rods $3\frac{1}{2}$ yards 1 foot $8\frac{1}{4}$ inches. Ans.

To avoid fractions as much as possible, we reduce the $\frac{1}{2}$ yard to inches, obtaining 18 inches; this added to $8\frac{1}{4}$ inches, gives $26\frac{1}{4}$ inches, which equals 2 feet $2\frac{1}{4}$ inches. Then, 2 feet $+ 1$ foot $= 3$ feet $= 1$ yard, and 3 yards $+ 1$ yard $= 4$ yards. Hence, the above answer becomes 11 rods 4 yards 0 feet $2\frac{1}{4}$ inches.

222. EXAMPLE.—What is the difference between 3 rd. 2 yd. 2 ft. 10 in. and 47 ft.?

SOLUTION.—47 ft. $= 2$ rd. 4 yd. 2 ft.

	rd.	yd.	ft.	in.
	3	2	2	10
	2	4	2	0
	0	$3\frac{1}{2}$	0	10
or		3	2	4 Ans.

To find (approximately) the interval of time between two dates:

223. EXAMPLE.—How many years, months, days, and hours between 4 o'clock P.M. of June 15, 1868, and 10 o'clock A.M., September 28, 1891?

SOLUTION.—

	yr.	mo.	da.	hr.
	1891	8	28	10
	1868	5	15	16
	23	3	12	18 Ans.

EXPLANATION.—Counting 24 hours in 1 day, 4 o'clock P.M. is the 16th hour from the beginning of the day, or midnight. On September 28, 8 months and 28 days have elapsed, and on June 15, 5 months and 15 days. After placing the earlier date under the later date, subtract as in the previous problems. Count 30 days as 1 month.

ARITHMETIC.

224. Rule 32.—*Place the smaller quantity under the larger quantity, with like denominations under each other. Beginning at the right, subtract successively the number in the subtrahend in each denomination from the one above, and place the differences underneath. If the number in the minuend of any denomination is less than the number under it in the subtrahend, one must be borrowed from the minuend of the next higher denomination, reduced, and added to it.*

EXAMPLES FOR PRACTICE.

225. From

(*a*) 125 lb. 8 oz. 14 pwt. 18 gr. take 96 lb. 9 oz. 10 pwt. 4 gr.
(*b*) 126 hhd. 27 gal. take 104 hhd. 14 gal. 1 qt. 1 pt.
(*c*) 65 T. 14 cwt. 64 lb. 10 oz. take 16 T. 11 cwt. 14 oz.
(*d*) 148 sq. yd. 16 sq. ft. 142 sq. in. take 132 sq. yd. 136 sq. in.
(*e*) 100 bu. take 28 bu. 2 pk. 5 qt. 1 pt.
(*f*) 14 mi. 34 rd. 16 yd. 13 ft. 11 in. take 3 mi. 27 rd. 11 yd. 4 ft. 10 in.

Ans.
(*a*) 28 lb. 11 oz. 4 pwt. 14 gr.
(*b*) 22 hhd. 12 gal. 2 qt. 1 pt.
(*c*) 49 T. 3 cwt. 63 lb. 12 oz.
(*d*) 16 sq. yd. 16 sq. ft. 6 sq. in.
(*e*) 71 bu. 1 pk. 2 qt. 1 pt.
(*f*) 11 mi. 7 rd. 5 yd. 9 ft. 1 in.

MULTIPLICATION OF DENOMINATE NUMBERS.

226. Example.—Multiply 7 lb. 5 oz. 13 pwt. 15 gr. by 12.

Solution.—

lb.	oz.	pwt.	gr.
7	5	13	15
			12
89	8	3	12 Ans.

Explanation.—15 grains × 12 = 180 grains. 180 ÷ 24 = 7 pennyweights and 12 grains remaining. Place the 12 in the grain column and carry the 7 pennyweights to the next. Now, 13 × 12 + 7 = 163 pennyweights; 163 ÷ 20 = 8 ounces and 3 pennyweights remaining. Then, 5 × 12 + 8 = 68 ounces; 68 ÷ 12 = 5 pounds and 8 ounces remaining. Then, 7 × 12 + 5 = 89 pounds. The entire product is 89 pounds 8 ounces 3 pennyweights 12 grains. Ans.

ARITHMETIC.

227. Rule 33.—*Multiply the number representing each denomination by the multiplier, and reduce each product to the next higher denomination, writing the remainders under each denomination, and carrying the quotient to the next, as in addition of denominate numbers.*

228. NOTE.—In multiplication and division of denominate numbers, it is sometimes easier to reduce the number to the lowest denomination given before multiplying or dividing, especially if the multiplier or divisor is a decimal. Thus, in the above example, had the multiplier been 1.2, the easiest way to multiply would have been to reduce the number to grains, then, multiply by 1.2, and reduce the product to higher denominations. For example, 7 lb. 5 oz. 13 pwt. 15 gr. = 43,047 gr. 43,047 × 1.2 = 51,656.4 gr. = 8 lb. 11 oz. 12 pwt. 8.4 gr. Also, 43,047 × 12 = 516,564 gr. = 89 lb. 8 oz. 3 pwt. 12 gr., as above. The student may use either method.

EXAMPLES FOR PRACTICE.

229. Multiply

(*a*) 15 cwt. 90 lb. by 5; (*b*) 12 yr. 10 mo. 4 wk. 3 da. by 14; (*c*) 11 mi. 145 rd. by 20; (*d*) 12 gal. 4 pt. by 9; (*e*) 8 cd. 76 cu. ft. by 15; (*f*) 4 hhd. 3 gal. 1 qt. 1 pt. by 12.

Ans. {
(*a*) 79 cwt. 50 lb.
(*b*) 180 yr. 11 mo. 2 wk.
(*c*) 229 mi. 20 rd.
(*d*) 112 gal. 2 qt.
(*e*) 128 cd. 116 cu. ft.
(*f*) 48 hhd. 40 gal. 2 qt.
}

DIVISION OF DENOMINATE NUMBERS.

230. EXAMPLE.—Divide 48 lb. 11 oz. 6 pwt. by 8.

SOLUTION.—

```
           lb.   oz.   pwt.   gr.
       8 ) 48    11     6      0
           ─────────────────────
           6 lb. 1 oz. 8 pwt. 6 gr.  Ans.
```

EXPLANATION.—After placing the quantities as above, proceed as follows: 8 is contained in 48 six times without a remainder. 8 is contained in 11 ounces once with 3 ounces remaining. 3 × 20 = 60; 60 + 6 = 66 pennyweights; 66 pennyweights ÷ 8 = 8 pennyweights and 2 remaining; 2 × 24 grains = 48 grains; 48 grains ÷ 8 = 6 grains. . Therefore, the entire quotient is 6 pounds 1 ounce 8 pennyweights 6 grains. Ans.

EXAMPLE.—A silversmith melted up 2 lb. 8 oz. 10 pwt. of silver, which he made into 6 spoons; what was the weight of each spoon?

ARITHMETIC.

SOLUTION.—

```
        lb.   oz.   pwt.
     6 ) 2    8     10
        ─────────────────
        5 oz.  8 pwt.  8 gr.  Ans.
```

EXPLANATION.—Since **we can not** divide 2 pounds by 6, we reduce it to ounces. 2 pounds = 24 ounces, and 24 ounces + 8 ounces = 32 ounces; 32 ounces ÷ 6 = 5 ounces and 2 ounces over. 2 ounces = 40 pennyweights. 40 pennyweights + 10 pennyweights = 50 pennyweights, and 50 pennyweights ÷ 6 = 8 pennyweights and 2 pennyweights over. 2 pennyweights = 48 grains, and 48 grains ÷ 6 = 8 grains. Hence, each spoon contains 5 ounces 8 pennyweights 8 grains. Ans.

ARITHMETIC. 75

231. EXAMPLE.—Divide 820 rd. 4 yd. 2 ft. by 112.

SOLUTION.—

```
                          rd.   yd. ft.   rd. yd. ft.  in.
              112 ) 8 2 0   4   2 ( 7   1   2   5.143  Ans.
                    7 8 4
                    ─────
                      3 6 rd. rem.
                      5.5
                    ─────
                    1 8 0
                    1 8 0
                    ─────
                    1 9 8.0 yd.
                          4
              112 ) 2 0 2 yd. ( 1 yd.
                    1 1 2
                    ─────
                      9 0 yd. rem.
                          3
                    ─────
                    2 7 0 ft.
                      2 ft.
              112 ) 2 7 2 ft. ( 2 ft.
                    2 2 4
                    ─────
                      4 8 ft. rem.
                      1 2
                    ─────
                      9 6
                      4 8
              112 ) 5 7 6 in. ( 5.1 4 2 8 + in., or 5.1 4 3 in.
                    5 6 0
                    ─────
                      1 6 0
                      1 1 2
                      ─────
                        4 8 0
                        4 4 8
                        ─────
                          3 2 0
                          2 2 4
                          ─────
                            9 6 0
                            8 9 6
                            ─────
                              6 4
```

EXPLANATION.—The first quotient is 7 rods with 36 rods remaining. $5.5 \times 36 = 198$ yards; 198 yards + 4 yards = 202 yards; 202 yards $\div 112 = 1$ yard and 90 yards remaining. $90 \times 3 = 270$ feet; 270 feet + 2 feet = 272 feet; 272 feet $\div 112 = 2$ feet and 48 feet remaining; $48 \times 12 = 576$ inches; 576 inches $\div 112 = 5.143$ inches, nearly. Ans.

ARITHMETIC.

The preceding example is solved by long division, because the numbers are too large to deal with mentally. Instead of expressing the last result as a decimal, it might have been expressed as a common fraction. Thus, $576 \div 112 = 5\frac{16}{112} = 5\frac{1}{7}$ inches. The chief advantage of using a common fraction is that if the quotient be multiplied by the divisor, the result will always be the same as the original dividend.

232. Rule 34.—*Find how many times the divisor is contained in the first or highest denomination of the dividend. Reduce the remainder (if any) to the next lower denomination, and add to it the number in the given dividend expressing that denomination. Divide this new dividend by the divisor. The quotient will be the next denomination in the quotient required. Continue in this manner until the lowest denomination is reached. The successive quotients will constitute the entire quotient.*

EXAMPLES FOR PRACTICE.

233. Divide

(*a*) 876 mi. 276 rd. by 22; (*b*) 1,137 bu. 3 pk. 4 qt. 1 pt. by 10; (*c*) 84 cwt. 48 lb. 49 oz. by 16; (*d*) 78 sq. yd. 18 sq. ft. 41 sq. in. by 18; (*e*) 148 mi. 64 rd. 24 yd. by 12; (*f*) 100 tons 16 cwt. 18 lb. 11 oz. by 15; (*g*) 36 lb. 18 oz. 18 pwt. 14 gr. by 8; (*h*) 112 mi. 48 rd. by 100.

Ans.
(*a*) 17 mi. $41\frac{7}{11}$ rd.
(*b*) 113 bu. 3 pk. 1 qt. $\frac{1}{4}$ pt.
(*c*) 5 cwt. 28 lb. $3\frac{1}{16}$ oz.
(*d*) 4 sq. yd. 4 sq. ft. $2\frac{5}{18}$ sq. in.
(*e*) 12 mi. 112 rd. 2 yd.
(*f*) 6 tons 14 cwt. 41 lb. $3\frac{11}{15}$ oz.
(*g*) 4 lb. 8 oz. 7 pwt. $7\frac{3}{4}$ gr.
(*h*) 1 mi. $38\frac{48}{100}$ rd.

INVOLUTION.

234. Involution is the process of multiplying a number by itself one or more times. The product obtained by multiplying a number by itself is called a **power** of that number.

Thus, the *second* power of 3 is 9, since 3×3 are 9.

ARITHMETIC. 77

The *third* power of 3 is 27, since $3 \times 3 \times 3$ are 27.
The *fifth* power of 2 is 32, since $2 \times 2 \times 2 \times 2 \times 2$ are 32.

235. An **exponent** is a *small figure* placed to the *right* and a little *above* a number to show to what *power* it is to be raised, or how many times the number is to be used as a factor, as the small figures ¹, ³, and ⁵ below:

$$3^2 = 3 \times 3 = 9.$$
$$3^3 = 3 \times 3 \times 3 = 27.$$
$$2^5 = 2 \times 2 \times 2 \times 2 \times 2 = 32.$$

236. The **root** of a number is that number which, used the required number of times as a factor, produces the number. In the above cases, 3 is a root of 9, since 3×3 are 9. It is also a root of 27, since $3 \times 3 \times 3$ are 27. Also, 2 is a root of 32, since $2 \times 2 \times 2 \times 2 \times 2$ are 32.

237. The *second* power of a number is called its **square**.

Thus, 5^2 is called the **square** of 5, or 5 *squared*, and its value is $5 \times 5 = 25$.

238. The *third* power of a number is called its **cube**.

Thus, 5^3 is called the *cube* of 5, or 5 *cubed*, and its value is $5 \times 5 \times 5 = 125$.

To find any power of a number:

239. EXAMPLE.—What is the third power, or cube, of 35?

SOLUTION.— $35 \times 35 \times 35$,

or

$$\begin{array}{r} 35 \\ 35 \\ \hline 175 \\ 105 \\ \hline 1225 \\ 35 \\ \hline 6125 \\ 3675 \\ \hline cube = 42875 \end{array}$$ Ans.

240. EXAMPLE.—What is the fourth power of 15?

SOLUTION.— $15 \times 15 \times 15 \times 15,$

or
```
      15
      15
      ──
      75
      15
     ───
     225
      15
    ────
    1125
     225
    ────
    3375
      15
   ─────
   16875
    3375
   ─────
```
fourth power = 50625 Ans.

241. EXAMPLE.— $1.2^3 =$ what?

SOLUTION.— $1.2 \times 1.2 \times 1.2,$

or
```
     1.2
     1.2
    ────
    1.44
     1.2
    ────
     288
     144
    ─────
   1.728  Ans.
```

242. EXAMPLE.—What is the third power, or cube, of $\frac{3}{8}$?

SOLUTION.— $\left(\tfrac{3}{8}\right)^3 = \tfrac{3}{8} \times \tfrac{3}{8} \times \tfrac{3}{8} = \dfrac{3 \times 3 \times 3}{8 \times 8 \times 8} = \tfrac{27}{512}.$ Ans.

243. Rule 35.—(**a**) *To raise a whole number or a decimal to any power, use it as a factor as many times as there are units in the exponent.*

(**b**) *To raise a fraction to any power, raise both the numerator and denominator to the power indicated by the exponent.*

EXAMPLES FOR PRACTICE.

244. Raise the following to the powers indicated:

(*a*) 85^2.
(*b*) $\left(\tfrac{11}{12}\right)^2$.
(*c*) 6.5^2.
(*d*) 14^4.

Ans.
(*a*) 7,225.
(*b*) $\tfrac{121}{144}$.
(*c*) 42.25.
(*d*) 38,416.

ARITHMETIC. 79

(e) $(\frac{2}{3})^2$.
(f) $(\frac{1}{4})^3$.
(g) $(\frac{2}{5})^2$.
(h) 1.4^5.

Ans. $\begin{cases} (e) & \frac{4}{9}. \\ (f) & \frac{1}{64}. \\ (g) & \frac{4}{25}. \\ (h) & 5.37824. \end{cases}$

EVOLUTION.

245. **Evolution** is the reverse of involution. It is the process of finding the root of a number which is considered as a power.

246. The **square root** of a number is that number which, when used twice as a factor, produces the number.

Thus, 2 is the square root of 4, since 2×2, or 2^2, $= 4$.

247. The **cube root** of a number is that number which, when used three times as a factor, produces the number.

Thus, 3 is the cube root of 27, since $3 \times 3 \times 3$, or 3^3, $= 27$.

248. The **radical sign** $\sqrt{}$, when placed before a number, indicates that some root of that number is to be found.

249. The **index** of the root is a *small figure* placed *over* and to the *left* of the *radical sign*, to show what root is to be found.

Thus, $\sqrt[2]{100}$ denotes the *square root* of 100.

$\sqrt[3]{125}$ denotes the *cube root* of 125.

$\sqrt[4]{256}$ denotes the *fourth root* of 256, and so on.

250. When the square root is to be extracted, the index is generally omitted. Thus, $\sqrt{100}$ indicates the square root of 100. Also, $\sqrt{225}$ indicates the square root of 225.

SQUARE ROOT.

251. The *largest* number that can be written with *one* figure is 9, and $9^2 = 81$; the *largest* number that can be written with *two* figures is 99, and $99^2 = 9,801$; with *three* figures 999, and $999^2 = 998,001$; with *four* figures 9,999, and $9,999^2 = 99,980,001$, etc.

In *each* of the above it will be noticed that the square of the number contains just *twice* as many figures as the number.

In order to find the square root of a number, the first step is to find how many figures there will be in the root. This is done by pointing off the number into *periods* of *two* figures each, *beginning at the right*. The number of periods will indicate the number of figures in the root.

Thus, the square root of 83,740,801 must contain 4 figures, since, pointing off the periods, we get 83'74'08'01, or 4 periods; consequently, there must be 4 figures in the root. In like manner, the square root of 50,625 must contain 3 figures, since there are (5'06'25) 3 periods.

252. EXAMPLE.—Find the square root of 31,505,769.

SOLUTION.—

```
                                              root.
           (a)    5          31'50'57'69 ( 5613  Ans.
                  5     (b)  25
           (d)  100     (c)    650
                  6            636
                106     (e)   1457
                  6           1121
               1120          33669
                  1          33669
               1121              0
                  1
              11220
                  3
              11223
```

EXPLANATION.—Pointing off into periods of two figures each, it is seen that there are four figures in the root. Now, find the largest single number whose square is less than or equal to 31, the first period. This is evidently 5, since $6^2 = 36$, which is greater than 31. Write it to the right, as in long division, and also to the left, as shown at (*a*). This is the first figure of the root. Now, multiply the 5 at (*a*) by the 5 in the root, and write the result under the first period, as shown at (*b*). Subtract, and obtain 6 as a remainder.

Bring down the next period 50, and annex it to the remainder 6, as shown at (*c*), which we call the **dividend**. Add the root already found to the 5 at (*a*), getting 10, and annex a cipher to this 10, thus making it 100, which we call the **trial divisor**. Divide the dividend (*c*) by the trial divisor (*d*), and obtain 6, which is *probably* the next figure of the root. Write 6 in the root, as shown, and also add it

ARITHMETIC. 81

to 100, the trial divisor, making it 106. This is called the **complete divisor.**

Multiply this by 6, the second figure in the root, and subtract the result from the dividend (*c*). The remainder is 14, to which annex the next period, making it 1,457, as shown at (*e*), which we call the **new dividend.** Add the second figure of the root to the trial divisor 106, and annex a cipher, thus getting 1,120. Dividing 1,457 by 1,120, we get 1 as the next figure of the root. Adding this last figure of the root to 1,120, multiplying the result by it, and subtracting from 1,457, the remainder is 336.

Annexing the next and last period, 69, the result is 33,669. Now, adding the last figure of the root to 1,121, and annexing a cipher as before, the result is 11,220. Dividing 33,669 by 11,220, the result is 3, the fourth figure in the root. Adding it to 11,220, and multiplying the sum by it, the result is 33,669. Subtracting, there is no remainder; hence, $\sqrt{31,505,769} = 5,613$. Ans.

253. The square of any number wholly decimal always contains twice as many figures as the number squared. For example, $.1^2 = .01$; $.13^2 = .0169$; $.751^2 = .564001$, etc.

254. It will also be noticed that the number squared is always less than the decimal. Hence, if it be required to find the square root of a decimal, and the decimal has not an even number of figures in it, annex a cipher. The best way to determine the number of figures in the root of a decimal is to begin at the decimal point, and, going towards the *right*, point off the decimal into periods of two figures each. Then, if the last period contains but one figure, annex a cipher.

255. EXAMPLE.—What is the square root of .000576 ?

```
                              root
SOLUTION.—        2        .00 05 76 ( .024   Ans.
                  2            4
                 ──            ───
                 40           176
                  4           176
                 ──           ───
                 44             0
```

82 ARITHMETIC.

EXPLANATION.—Beginning at the decimal point, and pointing off the number into periods of two figures each, it is seen that the first period is composed of ciphers; hence, the first figure of the root must be a cipher. The remaining portion of the solution should be perfectly clear from what has preceded.

256. If the number is not a perfect power, the root will consist of an interminable number of decimal places. The result may be carried to any required number of decimal places by annexing periods of two ciphers each to the number.

257. EXAMPLE.—What is the square root of 3? Find the result to five decimal places.

```
                                   root
SOLUTION.—  1           3.00'00'00'00'00(1.73205+   Ans.
            1           1
           ──           ──
           20            200
            7            189
           ──           ────
           27           1100
            7           1029
          ───           ────
          340            7100
            3            6924
          ───           ─────
          343           1760000
            3           1732025
         ────           ───────
         3460             27975
            2
         ────
         3462
            2
        ─────
        346400
             5
        ──────
        346405
```

EXPLANATION.—Annexing five periods of two ciphers each to the right of the decimal point, the first figure of the root is 1. To get the second figure, we find that, in dividing 200 by 20, it is 10. This is evidently too large.

Trying 9, we add 9 to 20, and multiply 29 by 9, the result is 261, a result which is considerably larger than 200; hence,

9 is too large. In the same way it is found that 8 is also too large. Trying 7, 7 times 27 is 189, a result smaller than 200; therefore, 7 is the second figure of the root. The next two figures, 3 and 2, are easily found. The fifth figure in the root is a cipher, since the trial divisor 34,640 is greater than the new dividend 17,600. In a case of this kind, we annex another cipher to 34,640, thereby making it 346,400, and bring down the next period, making the 17,600, 1,760,000. The next figure of the root is 5, and as we now have five decimal places, we will stop.

The square root of 3 is, then, 1.73205 +. Ans.

258. EXAMPLE.—What is the square root of .3 to five decimal places?

```
                                      root
SOLUTION.—     5           .30'00'00'00'00( .54772+  Ans.
               5            25
              ———           ———
             100            500
               4            416
              ———           ———
             104           8400
               4           7609
              ———          ————
            1080          79100
               7          76629
             ————         —————
            1087         247100
               7         219084
            —————        ——————
           10940          28016
               7
           —————
           10947
               7
           —————
          109540
               2
          ——————
          109542
```

EXPLANATION.—In the above example, we annex a cipher to .3, making the first period .30, since every period of a decimal, as was mentioned before, must have two figures in it. The remainder of the work should be perfectly clear.

259. If it is required to find the square root of a mixed number, begin at the decimal point, and point off the periods both ways. The manner of finding the root will then be exactly the same as in the previous cases.

260. EXAMPLE.—What is the square root of 258.2449?

```
SOLUTION.—      1         2'58.24'49(16.07  Ans.
                1         1
               20         158
                6         156
               26         22449
                6         22449
             3200             0
                7
             3207
```

EXPLANATION.—In the above example, since 320 is greater than 224, we place a cipher for the third figure of the root, and annex a cipher to 320, making it 3,200. Then, bringing down the next period 49, 7 is found to be the fourth figure of the root. Since there is no remainder, the square root of 258.2449 is 16.07. Ans.

261. Proof.—*To prove square root, square the result obtained. If the number is an exact power, the square of the root will equal it; if it is not an exact power, the square of the root will very nearly equal it.*

262. Rule 36.—(*a*) *Begin at units place, and separate the number into periods of two figures each, proceeding from left to right with the decimal part, if there is any.*

(*b*) *Find the greatest number whose square is contained in the first or left-hand period. Write this number as the first figure in the root; also, write it at the left of the given number.*

Multiply this number at the left by the first figure of the root, and subtract the result from the first period; then annex the second period to the remainder.

(*c*) *Add the first figure of the root to the number in the first column on the left, and annex a cipher to the result; this is the trial divisor. Divide the dividend by the trial divisor*

ARITHMETIC.

for the second figure in the root, and add this figure to the trial divisor to form the complete divisor. Multiply the complete divisor by the second figure in the root, and subtract this result from the dividend. (If this result is larger than the dividend, a smaller number must be tried for the second figure of the root.) Now bring down the third period, and annex it to the last remainder for a new dividend. Add the second figure of the root to the complete divisor, and annex a cipher for a new trial divisor.

(**d**) Continue in this manner to the last period, after which, if any additional places in the root are required, bring down cipher periods, and continue the operation.

(**e**) If at any time the trial divisor is not contained in the dividend, place a cipher in the root, annex a cipher to the trial divisor, and bring down another period.

(**f**) If the root contains an interminable decimal, and it is desired to terminate the operation at some point, say the fourth decimal place, carry the operation one place further, and if the fifth figure is 5 or greater, increase the fourth figure by 1 and omit the sign $+$.

263. Short Method.—If the number whose root is to be extracted is not an exact square, the root will be an interminable decimal. It is then usual to extract the root to a certain number of decimal places. In such cases, the work may be greatly shortened as follows: Determine to how many decimal places the work is to be carried, say 5, for example; add to this the number of places in the integral part of the root, say 2, for example, thus determining the number of figures in the root, in this case $5 + 2 = 7$. Divide this number by 2 and take the next higher number. In the above case, we have $7 \div 2 = 3\frac{1}{2}$; hence, we take 4, the next higher number. Now extract the root in the usual manner until the same number of figures have been obtained as was expressed by the number obtained above, in this case 4. Then form the trial divisor in the usual manner, but omitting to annex the cipher; divide the last remainder by the trial divisor, as in long division, obtaining as many figures of the quotient as there are remaining figures of the root, in this case $7 - 4 = 3$. The remainder so obtained is the remaining figures of the root.

Consider the example in Art. **258.** Here there are 5 figures in the root. We therefore extract the root to 3 places in the usual manner, obtaining .547 for the first three root figures. The next trial divisor is 1,094 (with the cipher omitted), and the last remainder is 791. Then, $791 \div 1,094 = .723$, and the next two figures of the root are 72, the whole root being $.54772 +$. Always carry the division one place further than desired, and if the last figure is 5 or greater, increase the preceding figure by 1. This method should not be used unless the root contains five or more figures.

NOTE.—If the last figure of the root found in the regular manner is a cipher, carry the process one place further before dividing as described above.

EXAMPLES FOR PRACTICE.

264. Find the square root of

(*a*) 186,624.			(*a*) 432.
(*b*) 2,050,624.			(*b*) 1,432.
(*c*) 29,855,296.			(*c*) 5,464.
(*d*) .0116964.			(*d*) .1081 +.
(*e*) 198.1369.			(*e*) 14.0761.
(*f*) 994,009.	Ans.		(*f*) 997.
(*g*) 2.375 to four decimal places.			(*g*) 1.5411.
(*h*) 1.625 to three decimal places.			(*h*) 1.275.
(*i*) .3025.			(*i*) .55.
(*j*) .571428.			(*j*) .7559 +.
(*k*) .78125.			(*k*) .8839.

CUBE ROOT.

265. In the same manner as in the case of square root, it can be shown that the periods into which a number is divided, whose cube root is to be extracted, must contain

ARITHMETIC.

three figures, except that the first or left-hand period of a whole or mixed number may contain one, two, or three figures.

266. EXAMPLE.—What is the cube root of 375,741,853,696?

SOLUTION.—

(1)	(2)	(3)	root
7	49	3 7 5'7 4 1'8 5 3'6 9 6 (7 2 1 6	Ans.
7	98	3 4 3	
14	14700	3 2 7 4 1	
7	4 2 4	3 0 2 4 8	
2 1 0	1 5 1 2 4	2 4 9 3 8 5 3	
2	4 2 8	1 5 5 7 3 6 1	
2 1 2	1 5 5 5 2 0 0	9 3 6 4 9 2 6 9 6	
2	2 1 6 1	9 3 6 4 9 2 6 9 6	
2 1 4	1 5 5 7 3 6 1	0	
2	2 1 6 2		
2 1 6 0	1 5 5 9 5 2 3 0 0		
1	1 2 9 8 1 6		
2 1 6 1	1 5 6 0 8 2 1 1 6		
1			
2 1 6 2			
1			
2 1 6 3 0			
6			
2 1 6 3 6			

EXPLANATION.—Write the work in three columns as follows: On the right place the number whose cube root is to be extracted, and point it off into periods of three figures each. Call this column (3). Find the largest number whose cube is less than or equal to the first period, in this case 7. Write the 7 on the right, as shown, for the first figure of the root, and also on the extreme left at the head of column (1). Multiply the 7 in column (1) by the first figure of the root 7, and write the product 49 at the head of column (2). Multiply the number in column (2) by the first figure of the root 7, and write the product 343 under the figures in the first period. Subtract and bring down the next period, obtaining 32,741 for the dividend. Add the first figure of the root to the number in column (1), obtaining 14, which call the *first correction*. Multiply the first correction by the first figure of the root, add the product to the number in column (2), and obtain 147. Add the first figure of the root to the first correction, and obtain 21, which call the *second correction*. Annex *two* ciphers to the number in column (2), and obtain 14,700 for the trial divisor; also, annex *one* cipher to the second correction, and obtain 210. Dividing the dividend by the trial divisor, we obtain $\frac{32,741}{14,700} = 2+$, and write the 2 as the second figure of the root. Add the 2 to the second correction, and obtain 212, which, multiplied by the second figure of the root, and added to the trial divisor, gives 15,124, the complete divisor. This last result, multiplied by the second figure of the root and subtracted from the dividend, gives a remainder of 2,493. Annexing the third period, we obtain 2,493,853 for the new dividend. Adding the second figure of the root to the number in column (1), we get 214 as the new first correction; this, multiplied by the second figure of the root and added to the trial divisor, gives 15,552. Adding the second figure of the root to the first new correction gives 216 as the second new correction. Annexing two ciphers to the number in column (2) gives 1,555,200, the new trial divisor. Annexing one cipher to the second new correction gives 2,160. Dividing the new dividend by the new trial divisor, we obtain $\frac{2,493,853}{1,555,200} = 1+$, and write 1 as the third figure of the root. The remainder of the work should be perfectly clear from what has preceded.

267. In extracting the cube root of a decimal, proceed as above, taking care that each period contains *three* figures. Begin the pointing off at the decimal point, going towards the right. If the last period does not contain three figures, annex ciphers until it does.

ARITHMETIC.

268. EXAMPLE.—What is the cube root of .009129329?

SOLUTION.—

```
                                            root
      2              4            .009'129'329(.209
      2              8             8
      ―              ―             ―――――――
      4           120000           1129329
      2             5481           1129329
      ―           ―――――――          ―――――――
     600          125481                 0
       9
      ―
     609
```

EXPLANATION.—Beginning at the decimal point, and pointing off as shown, the largest number whose cube is less than 9 is seen to be 2; hence, 2 is the first figure of the root. When finding the second figure, it is seen that the trial divisor 1,200 is greater than the dividend; hence, write a cipher for the second figure of the root; bring down the next period to form the new dividend; annex two ciphers to the trial divisor to form a new trial divisor; also, annex one cipher to the 60 in column (1). Dividing the new dividend by the new trial divisor, we get $\frac{1,129,329}{120,000} = 9\cdot +$, and write 9 as the third figure of the root. Complete the work as before.

269. EXAMPLE.—What is the cube root of 78,292.892952?

SOLUTION.—

```
                                                    root
      4              16           78'292.892'952 (42.78
      4              32            64
      ―              ――            ――――――
      8            4800            14292
      4             244            10088
     ――            ――――            ―――――――
    120            5044            4204892
      2             248            3766483
     ――            ――――――          ―――――――――
    122          529200            438409952
      2            8869            438409952
     ――            ――――――          ―――――――――
    124          538069                    0
      2            8918
     ――――         ――――――――
    1260         54698700
      7           102544
     ――――         ――――――――
    1267         54801244
      7
     ――――
    1274
      7
     ――――
    12810
       8
     ――――
    12818
```

ARITHMETIC.

EXPLANATION.—Since we have a mixed number, begin at the decimal point and point off periods of three figures each, in both directions. The first period contains but two figures, and the largest number whose cube is less than 78 is 4; consequently, 4 is the first figure of the root. The remainder of the work should be perfectly clear. When dividing the dividend by the trial divisor for the third figure of the root, the quotient was $8+$; but, on trying it, it was found that 8 was too large, the complete divisor being considerably larger than the trial divisor. Therefore, 7 was used instead of 8.

270. EXAMPLE.—What is the cube root of 5 to five decimal places?

SOLUTION.—

```
                                              root
   1           1        5.000'000'000'000'000(1.70997+
   1           2        1
   ─           ───       ─
   2          300        4000
   1          259        3913
   ─           ───       ────
  30          559          87000000
   7          308          78443829
   ──          ─────        ────────
  37         8670000        8556171000
   7           45981        7889992299
   ──          ──────       ──────────
  44         8715981        666178701000
   7           46062        614014317973
   ────        ───────       ────────────
  5100       876204300        52164383027
     9          461511
  ────        ─────────
  5109       876665811
     9          461592
  ────        ──────────
  5118      87712740300
     9          3590839
  ─────      ───────────
  51270     87716331139
      9
  ─────
  51279
      9
  ─────
  51288
      9
  ──────
  512970
       7
  ──────
  512977
```

ARITHMETIC. 91

EXPLANATION.—In the preceding example we annex five periods of ciphers, of three ciphers each, to the 5 for the decimal part of the root, placing the decimal point between the 5 and the first cipher. Since it is easy to see that the next figure of the root will be 5, we increase the last figure by 1, obtaining 1.70998 for the correct root to 5 decimal places. Ans.

271. EXAMPLE.—What is the cube root of .5 to four decimal places?

SOLUTION.—

```
                                            root
    7          49         .500'000'000'000(.7937+
    7          98.         343
   ──         ────        ─────
   14        14700        157000
    7         1971        150039
   ──         ────        ──────
  210        16671        6961000
    9         2052        5638257
  ───        ─────        ───────
  219      1872300       1322743000
    9         7119       1321748953
  ───      ───────       ──────────
  228      1879419          994047
    9         7128
  ───      ───────
 2370    188654700
    3       166579
 ────    ─────────
 2373    188821279
    3
 ────
 2376
    3
 ────
23790
    7
─────
23797
```

EXPLANATION.—In the above example we annex two ciphers to the .5 to complete the first period, and three periods of three ciphers each. The cube root of 500 is 7; this we write as the first figure of the root. The remainder of the work should be perfectly plain from the explanations of the preceding examples.

272. Example.—What is the cube root of .05 to four decimal places?

Solution.—

		root
3	9	.050'000'000'000(.3684+
3	18	27
6	2700	23000
3	576	19656
90	3276	3344000
6	612	3180032
96	388800	163968000
6	8704	162685504
102	397504	1282496
6	8768	
1080	40627200	
8	44176	
1088	40671376	
8		
1096		
8		
11040		
4		
11044		

273. Proof.—*To prove cube root, cube the result obtained. If the given number is an exact power, the cube of the root will equal it; if not an exact power, the cube of the root will very nearly equal it.*

274. Rule 37.—(*a*) *Arrange the work in three columns, placing the number whose cube root is to be extracted in the third or right-hand column. Begin at units place, and separate the number into periods of three figures each, proceeding from the decimal point towards the right with the decimal part, if there is any.*

(*b*) *Find the greatest number whose cube is not greater than the number in the first period. Write this number as the first figure of the root; also, write it at the head of the first column. Multiply the number in the first column by the first figure in the root, and write the result in the second*

ARITHMETIC.   93

column. *Multiply the number in the second column by the first figure of the root; subtract the product from the first period, and annex the second period to the remainder for a new dividend; add the first figure of the root to the number in the first column for the first correction. Multiply the first correction by the first figure of the root, and add the product to the number in the second column. Add the first figure of the root to the first correction to form the second correction. Annex one cipher to the second correction, and two ciphers to the last number in the second column; the last number in the second column is the trial divisor.*

(**c**) *Divide the dividend by the trial divisor to find the second figure of the root. Add the second figure of the root to the number in the first column, multiply the sum by the second figure of the root, and add the result to the trial divisor to form the complete divisor. Multiply the complete divisor by the second figure of the root, subtract the result from the dividend in the third column, and annex the third period to the remainder for a new dividend. Add the second figure of the root to the number in the first column to form the first correction; multiply the first correction by the second figure of the root, and add the product to the complete divisor. Add the second figure of the root to the first correction to form the second correction. Annex one cipher to the second correction, and two ciphers to the last number in the second column to form the new trial divisor.*

(**d**) *If there are more periods to be brought down, proceed as before. If there is a remainder after the root of the last period has been found, annex cipher periods, and proceed as before. The figures of the root thus obtained will be decimals.*

(**e**) *If the root contains an interminable decimal, and it is desired to terminate the operation at some point, say the fourth decimal place, carry the operation one place further, and if the fifth figure is 5 or greater, increase the fourth figure by 1 and omit the sign +.*

Art. **263** can be applied to cube root (or any other root) as well as to square root. Thus, in the example,

Art. **270**, there are to be $5+1=6$ figures in the root. Extracting the root in the usual manner to $6 \div 2 = 3$, say 4 figures, we get for the first four figures 1,709. The last remainder is 8,556,171, and the next trial divisor with the ciphers omitted is 8,762,043. Hence, the next two figures of the root are $8,556,171 \div 8,762,043 = .976$, say .98. Therefore, the root is 1.70998.

ROOTS OF FRACTIONS.

275. If a given number is in the form of a fraction and it is required to find some root of it, the simplest and most exact method is to reduce the fraction to a decimal and extract the required root of the decimal. If, however, the numerator and denominator of the fraction are perfect powers, extract the required root of each separately, and write the root of the numerator for a new numerator, and the root of the denominator for a new denominator.

276. EXAMPLE.—What is the square root of $\frac{9}{64}$?
SOLUTION.—$\sqrt{\frac{9}{64}} = \frac{\sqrt{9}}{\sqrt{64}} = \frac{3}{8}$.

277. EXAMPLE.—What is the square root of $\frac{5}{8}$?
SOLUTION.—$\sqrt{\frac{5}{8}} = \sqrt{.625} = .7906$.

278. EXAMPLE.—What is the cube root of $\frac{27}{64}$?
SOLUTION.—$\sqrt[3]{\frac{27}{64}} = \frac{\sqrt[3]{27}}{\sqrt[3]{64}} = \frac{3}{4}$.

279. EXAMPLE.—What is the cube root of $\frac{1}{4}$?
SOLUTION.—Since $\frac{1}{4} = .25$, $\sqrt[3]{\frac{1}{4}} = \sqrt[3]{.25} = .62996+$.

280. Rule 38.—*Extract the required root of the numerator and denominator separately; or, reduce the fraction to a decimal and extract the root of the decimal.*

EXAMPLES FOR PRACTICE.

281. Find the cube root of
(*a*) $\frac{27}{125}$.
(*b*) 2 to four decimal places.
(*c*) 4,180,769,192.462 to three decimal places.
(*d*) $\frac{27}{125}$.
(*e*) $\frac{3}{8}$.
(*f*) 513,229.783302144 to three decimal places.

Ans. $\begin{cases} (a)\ \frac{3}{5}. \\ (b)\ 1.2599+. \\ (c)\ 1,610.962. \\ (d)\ .8862+. \\ (e)\ .7211+. \\ (f)\ 80.064. \end{cases}$

ARITHMETIC.

TO EXTRACT OTHER ROOTS THAN THE SQUARE AND CUBE ROOTS.

282. EXAMPLE.—What is the fourth root of 256?

SOLUTION.—
$$\sqrt{256} = 16.$$
$$\sqrt{16} = 4.$$

Therefore, $\sqrt[4]{256} = 4.$ Ans.

In this example, $\sqrt[4]{256}$, the index is 4, which equals 2×2. The root indicated by 2 is the square root; therefore, the square root is extracted twice.

283. EXAMPLE.—What is the sixth root of 64?

SOLUTION.—
$$\sqrt{64} = 8.$$
$$\sqrt[3]{8} = 2.$$

Therefore, $\sqrt[6]{64} = 2.$ Ans.

In this example, $\sqrt[6]{64}$, the index is 6, which equals 2×3. The root indicated by 3 is the cube root; therefore, the square and cube roots are extracted in succession.

284. Rule 39.—*Separate the index of the required root into its factors (2's and 3's), and extract successively the roots indicated by the several factors obtained. The final result will be the required root.*

285. EXAMPLE.—What is the sixth root of 92,873,580 to two decimal places?

SOLUTION.— $6 = 3 \times 2$. Hence, extract the cube root, and then extract the square root of the result. $\sqrt[3]{92,873,580} = 452.8601$, and $\sqrt{452.8601} = 21.28 +$. Ans.

286. It matters not which root is extracted first, but it is probably easier and more exact to extract the cube root first.

EXAMPLES FOR PRACTICE.

287. Extract the

(a) Fourth root of 100.
(b) Fourth root of 3,049,800,625.
(c) Sixth root of 9,474,296,896.

Ans. { (a) 3.16227+.
(b) 235.
(c) 46.

RATIO.

288. Suppose that it is desired to compare two numbers, say 20 and 4. If we wish to know how many times larger 20 is than 4, we divide 20 by 4 and obtain 5 for the quotient; thus, $20 \div 4 = 5$. Hence, we say that 20 is 5 times as large as 4, i. e., 20 contains 5 times as many units as 4. Again, suppose we desire to know what part of 20 is 4. We then divide 4 by 20 and obtain $\frac{1}{5}$; thus, $4 \div 20 = \frac{1}{5}$, or .2. Hence, 4 is $\frac{1}{5}$ or .2 of 20. This operation of comparing two numbers is termed *finding the ratio* of the two numbers. Ratio, then, is a comparison. It is evident that the two numbers to be compared must be expressed in the same unit; in other words, the two numbers must both be abstract numbers or concrete numbers of the same kind. For example, it would be absurd to compare 20 horses with 4 birds, or 20 horses with 4. Hence, **ratio** may be defined as a comparison between two numbers of the same kind.

289. A ratio may be *expressed* in three ways; thus, if it is desired to compare 20 and 4, and express this comparison as a ratio, it may be done as follows: $20 \div 4$; $20 : 4$, or $\frac{20}{4}$. All three are read *the ratio of 20 to 4*. The ratio of 4 to 20 would be expressed thus: $4 \div 20$; $4 : 20$, or $\frac{4}{20}$. The first method of expressing a ratio, although correct, is seldom or never used; the second form is the one oftenest met with, while the third is rapidly growing in favor, and is likely to supersede the second. The third form, called the fractional form, is preferred by modern mathematicians, and possesses great advantages to students of algebra and of higher mathematical subjects. The second form seems to be better adapted to arithmetical subjects, and is one we shall ordinarily adopt. There is still another way of expressing a ratio, though seldom or never used in the case of a simple ratio like that given above. Instead of the colon, a straight vertical line is used; thus, 20 | 4.

ARITHMETIC.

290. The **terms** of a ratio are the two numbers to be compared; thus, in the above ratio, 20 and 4 are the terms. When both terms are considered together, they are called a **couplet**; when considered separately, the first term is called the **antecedent**, and the second term the **consequent**. Thus, in the ratio 20 : 4, 20 and 4 form a couplet, and 20 is the antecedent, and 4 the consequent.

291. A ratio may be **direct** or **inverse**. The *direct ratio* of 20 to 4 is 20 : 4, while the *inverse ratio* of 20 to 4 is 4 : 20. The direct ratio of 4 to 20 is 4 : 20, and the inverse ratio is 20 : 4. An inverse ratio is sometimes called a **reciprocal** ratio. The **reciprocal** of a number is 1 divided by the number. Thus, the reciprocal of 17 is $\frac{1}{17}$; of $\frac{3}{8}$ is $1 \div \frac{3}{8} = \frac{8}{3}$; i.e., the reciprocal of a fraction is the fraction inverted. Hence, the inverse ratio of 20 to 4 may be expressed as 4 : 20, or as $\frac{1}{20} : \frac{1}{4}$. Both have equal values; for, $4 \div 20 = \frac{1}{5}$, and $\frac{1}{20} \div \frac{1}{4} = \frac{1}{20} \times \frac{4}{1} = \frac{1}{5}$.

292. The term **vary** implies a ratio. When we say that two numbers vary as some other two numbers, we mean that the ratio between the first two numbers is the same as the ratio between the other two numbers.

293. The **value** of a ratio is the result obtained by performing the division indicated. Thus, the value of the ratio 20:4 is 5, it is the quotient obtained by dividing the antecedent by the consequent.

294. By expressing the ratio in the fractional form, for example, the ratio of 20 to 4 as $\frac{20}{4}$, it is easy to see, from the laws of fractions, that if both terms be multiplied, or both divided by the same number, it will not alter the value of the ratio. Thus,

$$\frac{20}{4} = \frac{20 \times 5}{4 \times 5} = \frac{100}{20}; \text{ and } \frac{20}{4} = \frac{20 \div 4}{4 \div 4} = \frac{5}{1}.$$

ARITHMETIC.

295. It is also evident, from the laws of fractions, that multiplying the antecedent or dividing the consequent multiplies the ratio; and dividing the antecedent or multiplying the consequent divides the ratio.

296. When a ratio is expressed in words, as the ratio of 20 to 4, the first number named is always regarded as the antecedent and the second as the consequent, without regard to whether the ratio itself is direct or inverse. *When not otherwise specified, all ratios are understood to be direct.* To express an inverse ratio, the simplest way of doing it is to express it as if it were a direct ratio, with the first number named as the antecedent, and then transpose the antecedent to the place occupied by the consequent and the consequent to the place occupied by the antecedent; or if expressed in the fractional form, invert the fraction. Thus, to express the inverse ratio of 20 to 4, first write it 20 : 4, and then, transposing the terms, as 4 : 20; or as $\frac{20}{4}$, and then inverting as $\frac{4}{20}$. Or, the reciprocals of the numbers may be taken, as explained above. To **invert** a ratio is to transpose its terms.

EXAMPLES FOR PRACTICE.

297. What is the value of the ratio of

(a) 98 to 49?
(b) $45 to $9?
(c) $6\frac{1}{4}$ to $\frac{1}{2}$?
(d) 3.5 to 4.5?
(e) The inverse ratio of 76 to 19?
(f) The inverse ratio of 49 to 98?
(g) The inverse ratio of 18 to 24?
(h) The inverse ratio of 9 to 15?
(i) The ratio of 10 to 3, multiplied by 3?
(j) The ratio of 35 to 49, multiplied by 7?
(k) The ratio of 18 to 64, divided by 9?
(l) The ratio of 14 to 28, divided by 5?

Ans.
(a) 2.
(b) 5.
(c) $12\frac{1}{4}$.
(d) $.77\frac{7}{9}$.
(e) $\frac{1}{4}$.
(f) 2.
(g) $1\frac{1}{3}$.
(h) $1\frac{2}{3}$.
(i) 10.
(j) 5.
(k) $\frac{1}{32}$.
(l) $\frac{1}{10}$.

298. Instead of expressing the value of a ratio by a single number, as above, it is customary to express it by

ARITHMETIC.

means of another ratio in which the consequent is 1. Thus, suppose that it is desired to find the ratio of the weights of two pieces of iron, one weighing 45 pounds and the other weighing 30 pounds. The ratio of the heavier to the lighter is then 45 : 30, an inconvenient expression. Using the fractional form, we have $\frac{45}{30}$. Dividing both terms by 30, the consequent, we obtain $\frac{1\frac{1}{2}}{1}$ or $1\frac{1}{2} : 1$. This is the same result as obtained above, for $1\frac{1}{2} \div 1 = 1\frac{1}{2}$, and $45 \div 30 = 1\frac{1}{2}$.

299. A ratio may be squared, cubed, or raised to any power, or any root of it may be taken. Thus, if the ratio of two numbers is 105 : 63, and it is desired to cube this ratio, the cube may be expressed as $105^3 : 63^3$. That this is correct is readily seen; for, expressing the ratio in the fractional form, it becomes $\frac{105}{63}$, and the cube is $\left(\frac{105}{63}\right)^3 = \frac{105^3}{63^3} = 105^3 : 63^3$. Also, if it is desired to extract the cube root of the ratio $105^3 : 63^3$, it may be done by simply dividing the exponents by 3, obtaining 105 : 63. This may be proved in the same way as in the case of cubing the ratio. Thus, $105^3 : 63^3 = \left(\frac{105}{63}\right)^3$, and $\sqrt[3]{\left(\frac{105}{63}\right)^3} = \frac{105}{63} = 105 : 63$.

300. Since $\left(\frac{105}{63}\right)^3 = \left(\frac{5}{3}\right)^3$, it follows that $105^3 : 63^3 = 5^3 : 3^3$ (this expression is read: the ratio of 105 cubed to 63 cubed equals the ratio of 5 cubed to 3 cubed), it follows that the antecedent and consequent may always be multiplied or divided by the same number, irrespective of any indicated powers or roots, without altering the value of the ratio. Thus, $24^2 : 18^2 = 4^2 : 3^2$. For, performing the operations indicated by the exponents, $24^2 = 576$ and $18^2 = 324$. Hence, $576 : 324 = 1\frac{7}{9}$ or $1\frac{7}{9} : 1$. Also, $4^2 = 16$ and $3^2 = 9$; hence, $16 : 9 = 1\frac{7}{9}$ or $1\frac{7}{9} : 1$, the same result as before. Also, $24^2 : 18^2 = \frac{24^2}{18^2} = \left(\frac{24}{18}\right)^2 = \left(\frac{4}{3}\right)^2 = \frac{4^2}{3^2} = 4^2 : 3^2$.

The statement may be proved for roots in the same manner. Thus, $\sqrt[3]{24^3} : \sqrt[3]{18^3} = \sqrt[3]{4^3} : \sqrt[3]{3^3}$. For the $\sqrt[3]{24^3} = 24$ and $\sqrt[3]{18^3} = 18$; and, $24 : 18 = 1\frac{1}{3}$ or $1\frac{1}{3} : 1$. Also, $\sqrt[3]{4^3} = 4$ and $\sqrt[3]{3^3} = 3$; $4 : 3 = 1\frac{1}{3}$ or $1\frac{1}{3} : 1$.

NOTE.—If the numbers composing the antecedent and consequent have different exponents, or if different roots of those numbers are indicated, the operations described in Art. **300** can not be performed. This is evident; for, consider the ratio $4^2 : 8^3$. When expressed in the fractional form, it becomes $\frac{4^2}{8^3}$, which can not be expressed either as $\left(\frac{4}{8}\right)^2$ or as $\left(\frac{4}{8}\right)^3$, and, hence, can not be reduced as described above.

PROPORTION.

301. Proportion is an equality of ratios, the equality being indicated by the double colon (::) or by the sign of equality (=). Thus, to write in the form of a proportion the two equal ratios, 8 : 4 and 6 : 3, which both have the same value 2, we may employ one of the three following forms:

$$8 : 4 :: 6 : 3 \qquad (1)$$
$$8 : 4 = 6 : 3 \qquad (2)$$
$$\frac{8}{4} = \frac{6}{3} \qquad (3)$$

302. The first form is the one most extensively used, by reason of its having been exclusively employed in all the older works on mathematics. The second and third forms are being adopted by all modern writers on mathematical subjects, and, in time, will probably entirely supersede the first form. In this paper we shall adopt the second form, unless some statement can be made clearer by using the third form.

303. A proportion may be *read* in two ways. The old way to read the above proportion was: *8 is to 4 as 6 is to 3;* the new way is: *the ratio of 8 to 4 equals the ratio of 6 to 3.* The student may read it either way, but we recommend the latter.

304. Each ratio of a proportion is termed a **couplet**. In the above proportion, 8 : 4 is a couplet, and so is 6 : 3.

ARITHMETIC. 101

305. The numbers forming the proportion are called **terms**; and they are numbered consecutively from left to right, thus:

$$\overset{\textit{first}}{8} : \overset{\textit{second}}{4} = \overset{\textit{third}}{6} : \overset{\textit{fourth}}{3}$$

Hence, in any proportion, the ratio of the first term to the second term equals the ratio of the third term to the fourth term.

306. The first and fourth terms of a proportion are called the **extremes**, and the second and third terms the **means**. Thus, in the foregoing proportion, 8 and 3 are the extremes and 4 and 6 are the means.

307. A **direct proportion** is one in which both couplets are direct ratios.

308. An **inverse proportion** is one which requires one of the couplets to be expressed as an inverse ratio. Thus, 8 is to 4 inversely as 3 is to 6 must be written $8 : 4 = 6 : 3$; i. e., the second ratio (couplet) must be inverted.

309. Proportion forms one of the most useful sections of arithmetic. In our grandfathers' arithmetics, it was called "The rule of three."

310. Rule 40.—*In any proportion, the product of the extremes equals the product of the means.*

Thus, in the proportion,

$$17 : 51 = 14 : 42.$$

$17 \times 42 = 51 \times 14$, since both products equal 714.

311. Rule 41.—*The product of the extremes divided by either mean gives the other mean.*

EXAMPLE.—What is the third term of the proportion $17 : 51 = : 42$?
SOLUTION.—Applying the rule, $17 \times 42 = 714$, and $714 \div 51 = 14$. Ans.

312. Rule 42.—*The product of the means divided by either extreme gives the other extreme.*

EXAMPLE.—What is the first term of the proportion $: 51 = 14 : 42$?
SOLUTION.—Applying the rule, $51 \times 14 = 714$, and $714 \div 42 = 17$. Ans.

313. When stating a proportion in which one of the terms is unknown, represent the missing term by a letter, as x. Thus, the last example would be written,

$$x : 51 = 14 : 42$$

and for the value of x we have $x = \dfrac{51 \times 14}{42} = 17$.

314. If the same (addition and subtraction excepted) operations be performed upon *all* of the terms of a proportion, the proportion is not thereby destroyed. In other words, if all of the terms of a proportion be (1) multiplied or (2) divided by the same number; (3) if all the terms be raised to the same power; if (4) the same root of all the terms be taken, or (5) if both couplets be inverted, the proportion still holds. We will prove these statements by a numerical example, and the student can satisfy himself by other similar ones. The fractional form will be used, as it is better suited to the purpose. Consider the proportion $8 : 4 = 6 : 3$. Expressing it in the third form, it becomes $\dfrac{8}{4} = \dfrac{6}{3}$. What we are to prove is that, if any of the five operations enumerated above be performed upon all of the terms of this proportion, the first fraction will still equal the second fraction.

1. Multiplying all the terms by any number, say 7, $\dfrac{8 \times 7}{4 \times 7} = \dfrac{6 \times 7}{3 \times 7}$; or $\dfrac{56}{28} = \dfrac{42}{21}$. Now $\dfrac{56}{28}$ evidently equals $\dfrac{42}{21}$, since the value of either ratio is 2, and the same is true of the original proportion.

2. Dividing all the terms by any number, say 7, $\dfrac{8 \div 7}{4 \div 7} = \dfrac{6 \div 7}{3 \div 7}$; or $\dfrac{8/7}{4/7} = \dfrac{6/7}{3/7}$. But $\dfrac{8}{7} \div \dfrac{4}{7} = 2$, and $\dfrac{6}{7} \div \dfrac{3}{7} = 2$ also, the same as in the original proportion.

3. Raising all the terms to the same power, say the cube, $\dfrac{8^3}{4^3} = \dfrac{6^3}{3^3}$. This is evidently true, since $\dfrac{8^3}{4^3} = \left(\dfrac{8}{4}\right)^3 = 2^3 = 8$, and $\dfrac{6^3}{3^3} = \left(\dfrac{6}{3}\right)^3 = 2^3 = 8$ also.

4. Extracting the same root of all the terms, say the cube root, $\frac{\sqrt[3]{8}}{\sqrt[3]{4}} = \frac{\sqrt[3]{6}}{\sqrt[3]{3}}$. It is evident that this is likewise true, since $\frac{\sqrt[3]{8}}{\sqrt[3]{4}} = \sqrt[3]{\frac{8}{4}} = \sqrt[3]{2}$, and $\frac{\sqrt[3]{6}}{\sqrt[3]{3}} = \sqrt[3]{\frac{6}{3}} = \sqrt[3]{2}$ also.

5. Inverting both couplets, $\frac{4}{8} = \frac{3}{6}$, which is true, since both equal $\frac{1}{2}$.

315. If both terms of either couplet be multiplied or both divided by the same number, the proportion is not destroyed. This should be evident from the preceding article, and also from Art. **294.** Hence, in any proportion, equal factors may be canceled from the terms of a couplet, before applying rule **41** or **42**. Thus, the proportion $45 : 9 = x : 7.1$, we may divide both terms of the first couplet by 9 (that is, cancel 9 from both terms), obtaining $5 : 1 = x : 7.1$, whence $x = 7.1 \times 5 \div 1 = 35.5$. (See note in Art. **300**.)

316. The principle of all calculations in proportion is this: *Three of the terms are always given, and the remaining one is to be found.*

317. EXAMPLE.—If 4 men can earn $25 in one week, how much can 12 men earn in the same time?

SOLUTION.—The required term must bear the same relation to the given term of the same kind as one of the remaining terms bears to the other remaining term. We can then form a proportion by which the required term may be found.

The first question the student must ask himself in every calculation by proportion is:
"What is it I want to find?"
In this case it is dollars. We have two sets of men, one set earning $25, and we want to know how many dollars the other set earns. It is evident that the *amount* 12 men earn bears the same relation to the *amount* that 4 men earn as 12 men bears to 4 men. Hence, we have the proportion, the amount 12 men earn is to $25 as 12 men is to 4 men; or, since either extreme equals the product of the means divided by the other extreme, we have

The amount 12 men earn : $25 = 12 men : 4 men,

or the amount 12 men earn $= \dfrac{\$25 \times 12}{4} = \$75$. Ans.

ARITHMETIC.

Since it matters not which place x or the required term occupies, the problem could be stated as any of the following forms, the value of x being the same in each:

(a) \$25 : the amount 12 men earn = 4 men : 12 men; or the amount 12 men earn $= \dfrac{\$25 \times 12}{4}$, or \$75, since either mean equals the product of the extremes divided by the other mean.

(b) 4 men : 12 men = \$25 : the amount 12 men earn; or the amount that 12 men earn $= \dfrac{\$25 \times 12}{4}$, or \$75, since either extreme equals the product of the means divided by the other extreme.

(c) 12 men : 4 men = the amount 12 men earn : \$25; or the amount that 12 men earn $= \dfrac{\$25 \times 12}{4}$, or \$75, since either mean equals the product of the extremes divided by the other mean.

318. If the proportion is an inverse one, first form it as though it were a direct proportion, and then invert one of the couplets.

EXAMPLES FOR PRACTICE.

319. Find the value of x in each of the following:

(a) \$16 : \$64 :: x : \$4.
(b) x : 85 :: 10 : 17.
(c) 24 : x :: 15 : 40.
(d) 18 : 94 :: 2 : x.
(e) \$75 : \$100 = x : 100.
(f) 15 pwt. : x = 21 : 10.
(g) x : 75 yd. = \$15 : \$5.

Ans.
(a) $x = \$1$.
(b) $x = 50$.
(c) $x = 64$.
(d) $x = 10\frac{4}{9}$.
(e) $x = 75$.
(f) $x = 7\frac{1}{7}$ pwt.
(g) $x = 225$ yd.

1. If 75 pounds of lead cost \$2.10, what would 125 pounds cost at the same rate? Ans. \$3.50.

2. If A does a piece of work in 4 days and B does it in 7 days, how long will it take A to do what B does in 63 days? Ans. 36 days.

3. The circumferences of any two circles are to each other as their diameters. If the circumference of a circle 7 inches in diameter is 22 inches, what will be the circumference of a circle 31 inches in diameter? Ans. $97\frac{3}{7}$ inches.

INVERSE PROPORTION.

320. In Art. **308,** an inverse proportion was defined as one which required one of the couplets to be expressed as an inverse ratio. Sometimes the word *inverse* occurs in the statement of the example; in such cases the proportion can be written directly, merely inverting one of the couplets. But it frequently happens that only by carefully studying the conditions of the example can it be ascertained whether the proportion is direct or inverse. When in doubt, the student can always satisfy himself as to whether the proportion is direct or inverse by first ascertaining what is required, and stating the proportion as a direct proportion. Then, in order that the proportion may be true, if the first term is smaller than the second term, the third term must be smaller than the fourth; or if the first term is larger than the second term, the third term must be larger than the fourth term. Keeping this in mind, the student can always tell whether the required term will be larger or smaller than the other term of the couplet to which the required term belongs. Having determined this, the student then refers to the example, and ascertains from its conditions whether the required term is to be larger or smaller than the other term of the same kind. If the two determinations agree, the proportion is direct; otherwise, it is inverse, and one of the couplets must be inverted.

321. EXAMPLE.—If A's *rate* of doing work is to B's as 5 : 7, and A does a piece of work in 42 days, in what time will B do it?

SOLUTION.—The required term is the number of days it will take B to do the work. Hence, stating as a direct proportion,

$$5 : 7 = 42 : x.$$

Now, since 7 is greater than 5, x will be greater than 42. But, referring to the statement of the example, it is easy to see that B works faster than A; hence it will take B a less number of days to do the work than A. Therefore, the proportion is an inverse one, and should be stated

$$5 : 7 = x : 42,$$

from which $x = \dfrac{5 \times 42}{7} = 30$ days. Ans.

Had the example been stated thus: The time that A requires to do a piece of work is to the time that B requires, as 5 : 7; A can do it in 42 days, in what time can B do it? it is evident that it would take B a longer time to do the work than it would A; hence, x would be greater than 42, and the proportion would be direct, the value of x being $\dfrac{7 \times 42}{5}$ = 58.8 days.

EXAMPLES FOR PRACTICE.

322. Solve the following:

1. If a pump which discharges 4 gal. of water per min. can fill a tank in 20 hr., how long will it take a pump discharging 12 gal. per min. to fill it? Ans. $6\frac{2}{3}$ hr.

2. The circular seam of a boiler requires 50 rivets when the pitch is $2\frac{1}{4}$ in.; how many would be required if the pitch were $3\frac{1}{8}$ in.? Ans. 40.

3. The spring hangers on a certain locomotive are $2\frac{1}{4}$ in. wide and $\frac{5}{8}$ in. thick; those on another engine are of same sectional area, but are 3 in. wide; how thick are they? Ans. $\frac{3}{8}$ in.

4. A locomotive with driving wheels 16 ft. in circumference runs a certain distance in 5,000 revolutions; how many revolutions would it make in going the same distance, if the wheels were 22 ft. in circumference (no allowance for slip being made in either case)? Ans. $3,636\frac{4}{11}$ rev.

POWERS AND ROOTS IN PROPORTION.

323. It was stated in Art. **299** that a ratio could be raised to any power or any root of it might be taken. A proportion is frequently stated in such a manner that one of the couplets must be raised to some power or some root of it must be taken. In all such cases, both terms of the couplet so affected *must be raised to the same power or the same root of both terms must be taken.*

324. EXAMPLE.—Knowing that the weight of a sphere varies as the cube of its diameter, what is the weight of a sphere 6 inches in diameter if a sphere 8 inches in diameter of the same material weighs 180 pounds?

SOLUTION.—This is evidently a direct proportion. Hence, we write

$$6^3 : 8^3 = x : 180.$$

Dividing both terms of the first couplet by 2^3 (see Art. **300**),

$$3^3 : 4^3 = x : 180, \text{ or } 27 : 64 = x : 180;$$

whence, $x = \dfrac{27 \times 180}{64} = 75\frac{15}{16}$ pounds. Ans.

EXAMPLE.—A sphere 8 inches in diameter weighs 180 pounds; what is the diameter of another sphere of the same material which weighs $75\frac{15}{16}$ pounds?

SOLUTION.—Since the weights of any two spheres are to each other as the cubes of their diameters, we have the proportion

$$180 : 75\tfrac{15}{16} = 8^3 : x^3.$$

ARITHMETIC. 107

x, the required term. must be cubed, because the other term of the couplet is cubed (see Art. **323**). But, $8^3 = 512$; hence,

$$180 : 75\tfrac{1\frac{5}{8}}{16} = 512 : x^3, \text{ or } x^3 = \frac{75\tfrac{1\frac{5}{8}}{16} \times 512}{180} = 216 \; ;$$

whence, $x = \sqrt[3]{216} = 6$ inches. Ans.

325. Since taking the same root of all of the terms of a proportion does not change its value (Art. **314**), the above example might have been solved by extracting the cube root of all of the numbers, thus obtaining $\sqrt[3]{180} : \sqrt[3]{75\tfrac{1\frac{5}{8}}{16}} = 8 : x$; whence,

$$x = \frac{8 \times \sqrt[3]{75\tfrac{1\frac{5}{8}}{16}}}{\sqrt[3]{180}} = 8 \times \sqrt[3]{\frac{75\tfrac{1\frac{5}{8}}{16}}{180}} = 8 \sqrt[3]{\frac{1,215}{2,880}} = 8 \sqrt[3]{\frac{27}{64}} =$$

$8 \times \tfrac{3}{4} = 6$ inches. The process, however, is longer and is not so direct, and the first method is to be preferred.

326. If two cylinders have *equal* volumes, but different diameters, the diameters are to each other inversely as the square roots of their lengths. Hence, if it is desired to find the diameter of a cylinder that is to be 15 inches long, and which shall have the same volume as one that is 9 inches in diameter and 12 inches long, we write the proportion

$$9 : x = \sqrt{15} : \sqrt{12}.$$

Since neither 12 nor 15 are perfect squares, we square all of the terms (Arts. **325** and **314**) and obtain

$$81 : x^2 = 15 : 12; \text{ whence } x^2 = \frac{81 \times 12}{15} = 64.8,$$

and $x = \sqrt{64.8} = 8.05$ inches $=$ diameter of 15-inch cylinder.

EXAMPLES FOR PRACTICE.

327. Solve the following examples:

1. The intensity of light varies inversely as the square of the distance from the source of light. If a gas jet illuminates an object 30 feet away with a certain distinctness, how much brighter will the object be at a distance of 20 feet ?   Ans. $2\tfrac{1}{4}$ times as bright.

2. In the last example, suppose that the object had been 40 feet from the gas jet; how bright would it have been compared with its brightness at 30 feet from the gas jet ?   Ans. $\tfrac{9}{16}$ as bright.

3. When comparing one light with another, the intensities of their illuminating powers vary as the squares of their distances from the

source. If a man can just distinguish the time indicated by his watch, 50 feet from a certain light, at what distance could he distinguish the time from a light 3 times as powerful? Ans. 86.6+ feet.

4. The quantity of air flowing through a mine varies directly as the square root of the pressure. If 60,000 cubic feet of air flow per minute when the pressure is 2.8 pounds per square foot, how much will flow when the pressure is 3.6 pounds per square foot?

Ans. 68,034 cu. ft. per min., nearly.

5. In the last example, suppose that 70,000 cubic feet per minute had been required; what would be the pressure necessary for this quantity? Ans. 3.81+ lb. per sq. ft.

CAUSES AND EFFECTS.

328. Many examples in proportion may be more easily solved by using the principle of *cause and effect*. That which may be regarded as producing a change or alteration in something, or as accomplishing something, may be called a **cause**, and the change, or alteration, or thing accomplished, is the **effect**.

329. *Like causes produce like effects.* Hence, when two causes of the same kind produce two effects of the same kind, the ratio of the causes equals the ratio of the effects; in other words, the first cause is to the second cause as the first effect is to the second effect. Thus, in the question, if 3 men can lift 1,400 pounds, how many pounds can 7 men lift? we call 3 men and 7 men the *causes* (since they accomplish something, viz., the lifting of the weight), the number of pounds lifted, viz., 1,400 pounds and x pounds, are the effects. If we call 3 men the first cause, 1,400 pounds is the first effect; 7 men is the second cause, and x pounds is the second effect. Hence, we may write

$$\text{1st cause} : \text{2d cause} = \text{1st effect} : \text{2d effect}$$
$$3 : 7 = 1,400 : x,$$

whence $x = \dfrac{7 \times 1,400}{3} = 3,266\tfrac{2}{3}$ pounds.

330. The principle of cause and effect is extremely useful in the solution of examples in compound proportion, as we shall now show.

ARITHMETIC. 109

COMPOUND PROPORTION.

331. All the cases of proportion so far considered have been cases of **simple proportion**; i. e., each term has been composed of but one number. There are many cases, however, in which two or all of the terms have more than one number in them; all such cases belong to **compound proportion**. In all examples in compound proportion, both causes or both effects or all four consist of more than two numbers. We will illustrate this by an

EXAMPLE.—If 40 men earn $1,280 in 16 days, how much will 36 men earn in 31 days?

SOLUTION.—Since 40 men earn something, 40 men is a cause, and since they take 16 days in which to earn something, 16 days is also a cause. For the same reason, 36 men and 31 days are also causes. The effects, that which is earned, are 1,280 dollars and x dollars. Then, 40 men and 16 days make up the first cause, and 36 men and 31 days make up the second cause. $1,280 is the first effect and x is the second effect. Hence, we write

$$\begin{array}{cccc} \textit{1st cause} & \textit{2d cause} & \textit{1st effect} & \textit{2d effect} \\ \begin{matrix}40\\16\end{matrix} & : \begin{matrix}36\\31\end{matrix} & = 1{,}280 & : x \end{array}$$

Now, instead of using the colon to express the ratio, we shall use the vertical line (see Art. 289), and the above becomes

$$\left.\begin{matrix}40\\16\end{matrix}\;\right|\;\begin{matrix}36\\31\end{matrix} = 1{,}280\;\left|\;x.\right.$$

In the last expression, the product of all of the numbers included between the vertical lines must equal the product of all the numbers without them; i. e., $36 \times 31 \times 1{,}280 = 40 \times 16 \times x$.

$$\text{Or } x = \frac{36 \times 31 \times \overset{\overset{2}{80}}{\cancel{1{,}280}}}{\cancel{40} \times \cancel{16}} = \$2{,}232. \quad \text{Ans.}$$

332. The above might have been solved by canceling factors of the numbers in the original proportion. For if any number within the lines has a factor common to any number without the lines, that factor may be canceled from both numbers. Thus, 16 is contained in

$$\left.\begin{matrix}\cancel{40}\\\cancel{16}\end{matrix}\;\right|\;\begin{matrix}36\\31\end{matrix} = \begin{matrix}\overset{2}{\cancel{80}}\\\cancel{1{,}280}\end{matrix}\;\left|\;x,\right.$$

1,280, 80 times. Cancel 16 and 1,280, and write 80 above 1,280. 40 is contained in 80, 2 times. Cancel 40 and 80, and write 2 above 80. Now, since there are no more numbers that can be canceled, $x = 36 \times 31 \times 2 = \$2,232$, he same result as was obtained in the last article.

333. Rule 43.—*Write all the numbers forming the first cause in a vertical column, and draw a vertical line; on the other side of this line write in a vertical column all of the numbers forming the second cause. Write the sign of equality to the right of the second column, and on the right of this form a third column of the numbers composing the first effect, drawing a vertical line to the right; on the other side of this line, write, for a fourth column, the numbers composing the second effect. There must be as many numbers in the second cause as in the first cause, and in the second effect as in the first effect; hence, if any term is wanting, write x in its place. Multiply together all of the numbers within the vertical lines, and also all those without the lines (canceling previously, if possible), and divide the product of those numbers which do not contain x by the product of the others in which x occurs, and the result will be the value of x.*

334. EXAMPLE.—If 40 men can dig a ditch 720 feet long, 5 feet wide and 4 feet deep in a certain time, how long a ditch 6 feet deep and 3 feet wide could 24 men dig in the same time?

SOLUTION.—Here 40 men and 24 men are the causes and the two ditches are the effects. Hence,

$$\begin{array}{c|c|c|c} & & 3 & \\ & & 18 & x \\ & & 720 & 3 \\ 40 & 24 & 5 & 3 \\ & & 4 & 6 \end{array}$$ whence, $x = 24 \times 5 \times 4 = 480$ feet. Ans.

335. EXAMPLE.—The volume of a cylinder varies directly as its length and directly as the square of its diameter. If the volume of a cylinder 10 inches in diameter and 20 inches long is 1,570.8 cubic inches, what is the volume of another cylinder 16 inches in diameter and 24 inches long?

SOLUTION.—In this example, either the dimensions or the volumes may be considered the causes; say we take the dimensions for the causes. Then, squaring the diameters,

$$\begin{array}{c|c|c|c|c|c} 10^2 & 16^2 & & 100 & 256 & \\ 20 & 24 & = 1{,}570.8 & x, \text{ or } & 20 & 24 & = 1{,}570.8 & x; \\ & & & & 5 & 6 & \end{array}$$

whence, $x = \dfrac{256 \times 6 \times 1{,}570.8}{5 \times 100} = 4{,}825.4976$ cubic inches. Ans.

ARITHMETIC.

336. EXAMPLE.—If a block of granite 8 ft. long, 5 ft. wide and 3 ft. thick weighs 7,200 lb., what will be the weight of a block of granite 12 ft. long, 8 ft. wide and 5 ft. thick?

SOLUTION.—Taking the weights as the effects, we have

$$\begin{array}{c|c|c} \cancel{8} & 4 \\ \cancel{5} & \cancel{12} \\ \cancel{5} & \cancel{8} = 7{,}200 \\ \cancel{3} & \cancel{5} \end{array} \quad x, \text{ or } x = 4 \times 7{,}200 = 28{,}800 \text{ pounds. Ans.}$$

337. EXAMPLE.—If 12 compositors in 30 days of 10 hours each set up 25 sheets of 16 pages each, 32 lines to the page, in how many days 8 hours long can 18 compositors set up, in the same type, 64 sheets of 12 pages each, 40 lines to the page?

SOLUTION.—Here compositors, days, and hours compose the causes, and sheets, pages, and lines the effects. Hence,

$$\begin{array}{c|c|c|c} 3 & \cancel{3} & \cancel{5} & 2 \\ \cancel{12} & \cancel{18} & \cancel{25} & \cancel{64} \\ \cancel{30} & x & = \cancel{16} & \cancel{12}, \text{ or } x = 3 \times 10 \times 2 = 60 \text{ days. Ans.} \\ \cancel{8} & & \cancel{4} \\ 10 & \cancel{8} & \cancel{32} & \cancel{40} \\ & & & \cancel{5} \end{array}$$

338. In examples stated like that in Art. **335,** should an inverse proportion occur, write the various numbers as in the preceding examples, and then transpose those numbers which are said to vary inversely from one side of the vertical line to the other side.

EXAMPLE.—The centrifugal force of a revolving body varies directly as its weight, as the square of its velocity and inversely as the radius of the circle described by the center of the body. If the centrifugal force of a body weighing 15 pounds is 187 pounds when the body revolves in a circle having a radius of 12 inches, with a velocity of 20 feet per second, what will be the centrifugal force of the same body when the radius is increased to 18 inches and the speed is increased to 24 feet per second?

SOLUTION.—Calling the centrifugal force the effect, we have,

$$\begin{array}{c|c|c|c} 15 & 15 & & \\ 20^2 & 24^2 & = & 187 & x. \\ 12 & 18 & & \end{array}$$

Transposing 12 and 18 (since the radii are to vary inversely) and squaring 20 and 24,

$$\begin{array}{c|c} \cancel{15} & \cancel{15} \\ & 2 \\ 25 & \cancel{36} = 187 \\ \cancel{400} & \cancel{576} \\ \cancel{18} & 12 \end{array} \quad x, \text{ or } x = \frac{12 \times 2 \times 187}{25} = 179.52 \text{ pounds. Ans.}$$

ARITHMETIC.

EXAMPLES FOR PRACTICE.

339. Solve the following by compound proportion:

1. If 12 men dig a trench 40 rods long in 24 days of 10 hours each, how many rods can 16 men dig in 18 days of 9 hours each?

Ans. 36 rods.

2. If a piece of iron 7 ft. long, 4 in. wide, and 6 in. thick weighs 600 lb., how much will a piece of iron weigh that is 16 ft. long, 8 in. wide and 4 in. thick? Ans. 1,828$\tfrac{4}{7}$ lb.

3. If 24 men can build a wall 72 rods long, 6 feet wide, and 5 feet high in 60 days of 10 hours each, how many days will it take 32 men to build a wall 96 rods long, 4 feet wide and 8 feet high, working 8 hours a day? Ans. 80 days.

4. The horsepower of an engine varies as the mean effective pressure, as the piston speed and as the square of the diameter of the cylinder. If an engine having a cylinder 14 inches in diameter develops 112 horsepower when the mean effective pressure is 48 pounds per square inch and the piston speed is 500 feet per minute, what horsepower will another engine develop if the cylinder is 16 inches in diameter, piston speed is 600 feet per minute, and mean effective pressure is 56 pounds per square inch? Ans. 204.8 horsepower.

5. Referring to the example in Art. **335,** what will be the volume of a cylinder 20 inches in diameter and 24 inches long?

Ans. 7,539.84 cubic inches.

6. Knowing that the product of $3 \times 5 \times 7 \times 9$ is 945, what is the product of $6 \times 15 \times 14 \times 36$? Ans. 45,360.

7. The speed in miles per hour of a locomotive is directly proportional to the diameter of its driving wheels and the number of revolutions they make in one minute. A locomotive with driving wheels 66 inches in diameter runs 29.45 miles in an hour when the wheels make 150 revolutions per minute; how many miles will be run in one hour by a locomotive having wheels 72 inches in diameter running 220 revolutions per minute? Ans. 47.12 miles per hour.

8. The capacity of a cylindrical boiler is proportional to its length and the square of its diameter. A boiler 12 feet long and 48 inches in diameter will hold 1,128 gallons; what is the capacity of a boiler 16 feet long and 42 inches in diameter? Ans. 1,151.5 gallons.

9. The power that may be transmitted by a belt is proportional to its width and the diameter and number of revolutions made by the pulley on which it runs. If a belt 12 inches wide will transmit 10 horsepower when running over a pulley 20 inches in diameter that makes 125 revolutions per minute, how many horsepower may be transmitted by a belt 8 inches wide when running over a pulley 30 inches in diameter that makes 200 revolutions per minute?

Ans. 16 horsepower.

10. The load that a beam supported at the two ends will carry is directly proportional to its width and the square of its depth, and inversely proportional to its length. If an oak beam 8 inches wide, 12 inches deep, and 15 feet long will safely carry a load of 13,824 pounds, what is the safe load that a beam 10 inches wide, 16 inches deep, and 20 feet long will support? Ans. 23,040 pounds.

Mensuration and Use of Letters in Algebraic Formulas.

FORMULAS.

340. The term **formula,** as used in mathematics and in technical books, may be defined as *a rule in which symbols are used instead of words;* in fact, a formula may be regarded as a shorthand method of expressing a rule. Any formula can be expressed in words, and when so expressed it becomes a rule.

Formulas are much more convenient than rules; they show at a glance all the operations that are to be performed; they do not require to be read three or four times, as is the case with most rules, to enable one to understand their meaning; they take up much less space, both in the printed book and in one's note-book, than rules; in short, whenever a rule can be expressed as a formula, the formula is to be preferred.

As the term "quantity" is a very convenient one to use, we will define it. In mathematics, the word **quantity** is applied to anything that it is desired to subject to the ordinary operations of addition, subtraction, multiplication, etc., when we do not wish to be more specific and state exactly what the thing is. Thus, we can say "two or more numbers," or "two or more quantities;" the word quantity is more general in its meaning than the word number.

341. The signs used in formulas are the ordinary signs indicative of operations, and the signs of aggregation. All these signs are explained in arithmetic, but some of them will here be explained in order to refresh the student's memory.

116 MENSURATION.

The signs indicative of operations are six in number, viz.: $+, -, \times, \div, |, \sqrt{}$.

Division is indicated by the sign $\div$, or by placing a straight line between the two quantities. Thus, $25 \mid 17$, $25 / 17$, and $\tfrac{25}{17}$ all indicate that 25 is to be divided by 17. When both quantities are placed on the same horizontal line, the straight line indicates that the quantity on the left is to be divided by that on the right. When one quantity is below the other, the straight line between indicates that the quantity above the line is to be divided by the one below it.

The sign ($\sqrt{}$) indicates that some root of the quantity to the right is to be taken; it is called the **radical sign**. To indicate what root is to be taken, a small figure, called the **index**, is placed within the sign, this being always omitted when the square root is to be indicated. Thus, $\sqrt{25}$ indicates that the square root of 25 is to be taken; $\sqrt[3]{25}$ indicates that the cube root of 25 is to be taken; etc.

The signs of aggregation are four in number; viz., ——, ( ), [ ], and { }, respectively called the **vinculum**, the **parenthesis**, the **brackets**, and the **brace**; they are used when it is desired to indicate that all the quantities included by them are to be subjected to the same operation. Thus, if we desire to indicate that the sum of 5 and 8 is to be multiplied by 7, and we do not wish to actually add 5 and 8 before indicating the multiplication, we may employ any one of the four signs of aggregation as here shown: $\overline{5+8} \times 7$, $(5+8) \times 7$, $[5+8] \times 7$, $\{5+8\} \times 7$. The vinculum is placed above those quantities which are to be treated as one quantity and subjected to the same operations.

While any one of the four signs may be used as shown above, custom has restricted their use somewhat. The vinculum is rarely used except in connection with the radical sign. Thus, instead of writing $\sqrt[3]{}(5+8)$, $\sqrt[3]{}[5+8]$, or $\sqrt[3]{}\{5+8\}$ for the cube root of 5 plus 8, all of which would be correct, the vinculum is nearly always used, $\sqrt[3]{5+8}$.

In cases where but one sign of aggregation is needed (except, of course, when a root is to be indicated), the parenthesis

MENSURATION.

is always used. Hence, $(5 + 8) \times 7$ would be the usual way of expressing the product of 5 plus 8, and 7.

If two signs of aggregation are needed, the brackets and parenthesis are used, so as to avoid having a parenthesis within a parenthesis, the brackets being placed outside. For example, $[(20 - 5) \div 3] \times 9$ means that the difference between 20 and 5 is to be divided by 3, and this result multiplied by 9.

If three signs of aggregation are required, the brace, brackets, and parenthesis are used, the brace being placed outside, the brackets next, and the parenthesis inside. For example, $\{[(20 - 5) \div 3] \times 9 - 21\} \div 8$ means that the quotient obtained by dividing the difference between 20 and 5 by 3 is to be multiplied by 9, and that after 21 has been subtracted from the product thus obtained, the result is to be divided by 8.

Should it be necessary to use all four of the signs of aggregation, the brace would be put outside, the brackets next, the parenthesis next, and the vinculum inside. For example, $\{[(\overline{20 - 5} \div 3) \times 9 - 21] \div 8\} \times 12$.

As stated in arithmetic, when several quantities are connected by the various signs indicating addition, subtraction, multiplication, and division, the operation indicated by the sign of multiplication must always be performed first. Thus, $2 + 3 \times 4$ equals 14, 3 being multiplied by 4, before adding to 2. Similarly, $10 \div 2 \times 5$ equals 1, since 2×5 equals 10, and $10 \div 10$ equals 1. Hence, in the above case, if the brace were omitted, the result would be $\frac{1}{4}$, whereas, by inserting the brace, the result is 36.

Following the sign of multiplication comes the sign of division in order of importance. For example, $5 - 9 \div 3$ equals 2, 9 being divided by 3 before subtracting from 5. The signs of addition and subtraction are of equal value; that is, if several quantities are connected by plus and minus signs, the indicated operations may be performed in the order in which the quantities are placed.

There is one other sign used, which is neither a sign of aggregation nor a sign indicative of an operation to be

MENSURATION.

performed; it is (=), and is called the sign of *equality*; it means that all on one side of it is exactly equal to all on the other side. For example, $2 = 2$, $5 - 3 = 2$, $5 \times (14 - 9) = 25$.

342. Having called particular attention to certain signs used in formulas, the formulas themselves will now be explained. First, consider the well-known rule for finding the horsepower of a steam-engine, which may be stated as follows:

Divide the continued product of the mean effective pressure in pounds per square inch, the length of the stroke in feet, the area of the piston in square inches, and the number of strokes per minute, by 33,000; the result will be the horsepower.

This is a very simple rule, and very little, if anything, will be saved by expressing it as a formula, so far as clearness is concerned. The formula, however, will occupy a great deal less space, as we shall show.

An examination of the rule will show that four quantities (viz., the mean effective pressure, the length of the stroke, the area of the piston, and the number of strokes) are multiplied together, and the result is divided by 33,000. Hence, the rule might be expressed as follows:

$$\text{Horsepower} = \frac{\text{mean effective pressure}}{\text{(in pounds per square inch)}} \times \frac{\text{stroke}}{\text{(in feet)}} \times \frac{\text{area of piston}}{\text{(in square inches)}} \times \frac{\text{number of strokes}}{\text{(per minute)}} \div 33{,}000.$$

This expression could be shortened by representing each quantity by a single letter; thus, representing horsepower by the letter "H," the mean effective pressure in pounds per square inch by "P," the length of stroke in feet by "L," the area of the piston in square inches by "A," the number of strokes per minute by "N," and substituting these letters for the quantities that they represent, the above expression would reduce to

$$H = \frac{P \times L \times A \times N}{33{,}000},$$

a much simpler and shorter expression. This last expression is called a *formula*.

The formula just given shows, as we stated in the beginning, that a formula is really a shorthand method of expressing a rule. It is customary, however, to omit the sign of multiplication between two or more quantities when they are to be multiplied together, or between a number and a letter representing a quantity, it being always understood that, when two letters are adjacent with no sign between them, the quantities represented by these letters are to be multiplied. Bearing this fact in mind, the formula just given can be further simplified to

$$H = \frac{PLAN}{33,000}.$$

The sign of multiplication, evidently, can not be omitted between two or more numbers, as it would then be impossible to distinguish the numbers. A near approach to this, however, may be attained by placing a dot between the numbers which are to be multiplied together, and this is frequently done in works on mathematics when it is desired to economize space. In such cases it is usual to put the dot higher than the position occupied by the decimal point. Thus, $2 \cdot 3$ means the same as 2×3; $542 \cdot 749 \cdot 1,006$ indicates that the numbers 542, 749, and 1,006 are to be multiplied together.

It is also customary to omit the sign of multiplication in expressions similar to the following: $a \times \sqrt{b+c}$, $3 \times (b+c)$, $(b+c) \times a$, etc., writing them $a\sqrt{b+c}$, $3(b+c)$, $(b+c)a$, etc. The sign is not omitted when several quantities are included by a vinculum, and it is desired to indicate that the quantities so included are to be multiplied by another quantity. For example, $3 \times \overline{b+c}$, $\overline{b+c} \times a$, $\sqrt{b+c} \times a$, etc., are always written as here printed.

343. Before proceeding further, we will explain one other device that is used by formula makers and which is apt to puzzle one who encounters it for the first time—it is the use of what mathematicians call *primes* and *subs.*, and what printers call *superior* and *inferior* characters. As a rule, formula makers designate quantities by the initial letters of the names of the quantities. For example, they represent volume by v, pressure by p, height by h, etc. This practice is to be commended, as the letter itself serves in many cases to identify the quantity which it represents. Some authors carry the practice a little further, and represent all quantities of the same nature by the same letter throughout the book, always having the same letter represent the same thing. Now, this practice necessitates the use of the primes and subs. above mentioned, when two quantities have the same name but represent different things. Thus, consider the word *pressure* as applied to steam, at different stages between the boiler and the condenser. First, there is *absolute* pressure, which is equal to the gauge pressure in pounds per square inch plus the pressure indicated by the barometer reading (usually assumed in practice to be 14.7 pounds per square inch, when a barometer is not at hand). If this be represented by p, how shall we represent the gauge pressure? Since the absolute pressure is always greater than the gauge pressure, suppose we decide to represent it by a capital letter, and the gauge pressure by a small (lower-case) letter. Doing so, P represents absolute pressure, and p, gauge pressure. Further, there is usually a "drop" in pressure between the boiler and the engine, so that the initial pressure, or pressure at the beginning of the stroke, is less than the pressure at the boiler. How shall we represent the initial pressure? We may do this in one of three ways and still retain the letter p or P to represent the word pressure: First, by the use of the prime mark; thus, p' or P' (read p *prime* and P *major prime*) may be considered to represent the initial gauge pressure, or the initial absolute pressure. Second, by the use of sub. figures; thus, p_1 or P_1 (read p *sub. one* and P *major sub. one*). Third, by the use of sub. letters; thus, p_i or P_i (read p *sub. i* and P *major sub. i*). In the same manner p'' (read p *second*), p_2, or p_r might be used to represent the gauge pressure at release, etc. The sub. letters have the advantage of still further identifying the quantity represented; in many instances, however, it is not convenient to use them, in which case primes and subs. are used instead. The prime notation may be continued as follows: p''', p^{iv}, p'', etc.; it is inadvisable to use superior figures, for example, p^1, p^2, p^3, p^a, etc., as they are liable to be mistaken for exponents.

The main thing to be remembered by the student is that *when a formula is given in which the same letters occur several times, all like letters having the same primes or subs. represent the same quantities, while those which differ in any respect represent different quantities.* Thus, in the formula

$$t = \frac{w_1 s_1 t_1 + w_2 s_2 t_2 + w_3 s_3 t_3}{w_1 s_1 + w_2 s_2 + w_3 s_3},$$

w_1, w_2, and w_3 represent the weights of three different bodies; s_1, s_2, and s_3, their specific heats; and t_1, t_2, and t_3, their temperatures; while t represents the final temperature after the bodies have been mixed together. It should be noted that those letters having the *same* subs. refer to the same bodies. Thus, w_1, s_1, and t_1 all refer to one of the three bodies; w_2, s_2, t_2, to another body; etc.

It is very easy to apply the above formula when the values of the quantities represented by the different letters are known. All that is required is to substitute the numerical values of the letters, and then perform the indicated operations. Thus, suppose that the values of w_1, s_1, and t_1 are, respectively, 2 pounds, .0951, and 80°; of w_2, s_2, and t_2, 7.8 pounds, 1, and 80°; and of w_3, s_3, and t_3, $3\tfrac{1}{4}$ pounds, .1138, and 780°; then, the final temperature t is, substituting these values for their respective letters in the formula,

$$t = \frac{2 \times .0951 \times 80 + 7.8 \times 1 \times 80 + 3\tfrac{1}{4} \times .1138 \times 780}{2 \times .0951 + 7.8 \times 1 + 3\tfrac{1}{4} \times .1138} =$$

$$\frac{15.216 + 624 + 288.483}{.1902 + 7.8 + .36985} = \frac{927.699}{8.36005} = 110.97°.$$

In substituting the numerical values, the signs of multiplication are, of course, written in their proper places; all the multiplications are performed before adding, according to the rule previously given.

MENSURATION.

344. The student should now be able to apply any formula involving only algebraic expressions that he may meet with, and which does not require the use of logarithms for its solution. We will, however, call his attention to one or two other facts that he may have forgotten.

Expressions similar to $\dfrac{160}{\dfrac{660}{25}}$ sometimes occur, the heavy line indicating that 160 is to be divided by the quotient obtained by dividing 660 by 25. If both lines were light it would be impossible to tell whether 160 was to be divided by $\dfrac{660}{25}$, or whether $\dfrac{160}{660}$ was to be divided by 25. If this latter result were desired, the expression would be written $\dfrac{\dfrac{160}{660}}{25}$. In every case the heavy line indicates that all above it is to be divided by all below it.

In an expression like the following, $\dfrac{160}{7+\dfrac{660}{25}}$, the heavy line is not necessary, since it is impossible to mistake the operation that is required to be performed. But, since $7+\dfrac{660}{25}=\dfrac{175+660}{25}$, if we substitute $\dfrac{175+660}{25}$ for $7+\dfrac{660}{25}$, the heavy line becomes necessary in order to make the resulting expression clear. Thus,

$$\dfrac{160}{7+\dfrac{660}{25}} = \dfrac{160}{\dfrac{175+660}{25}} = \dfrac{160}{\dfrac{835}{25}}$$

Fractional exponents are sometimes used instead of the radical sign. That is, instead of indicating the square, cube, fourth root, etc., of some quantity, as 37, by $\sqrt{37}$, $\sqrt[3]{37}$, $\sqrt[4]{37}$, etc., these roots are indicated by $37^{\frac{1}{2}}$, $37^{\frac{1}{3}}$, $37^{\frac{1}{4}}$, etc. Should the numerator of the fractional exponent be some quantity other than 1, this quantity, whatever it may be, indicates that the quantity affected by the exponent is to be raised to the power indicated by the numerator; the denominator is *always* the index of the root. Hence, instead of writing $\sqrt[3]{37^2}$ for the cube root of the square of 37, it may be written $37^{\frac{2}{3}}$, the denominator being the index of the root; in other words, $\sqrt[3]{37^2} = 37^{\frac{2}{3}}$. Likewise, $\sqrt[5]{(1+a^2 b)^2}$ may also be written $(1 + a^2 b)^{\frac{2}{5}}$, a much simpler expression.

345. We will now give several examples showing how to apply some of the more difficult formulas that the student may encounter.

1. The area of any segment of a circle that is less than (or equal to) a semicircle is expressed by the formula

$$A = \frac{\pi r^2 E}{360} - \frac{c}{2}(r-h),$$

in which A = area of segment;
$\pi = 3.1416$;
r = radius;
E = angle obtained by drawing lines from the center to the extremities of arc of segment;
c = chord of segment;
and h = height of segment.

EXAMPLE.—What is the area of a segment whose chord is 10 inches long, angle subtended by chord is 83.46°, radius is 7.5 inches, and height of segment is 1.91 inches?

SOLUTION.—Applying the formula just given,

$$A = \frac{\pi r^2 E}{360} - \frac{c}{2}(r-h) = \frac{3.1416 \times 7.5^2 \times 83.46}{360} - \frac{10}{2}(7.5 - 1.91) =$$
$40.968 - 27.95 = 13.018$ square inches, nearly. Ans.

2. The area of any triangle may be found by means of the following formula, in which A = the area, and a, b, and c represent the lengths of the sides:

$$A = \frac{b}{2}\sqrt{a^2 - \left(\frac{a^2+b^2-c^2}{2b}\right)^2}.$$

EXAMPLE.—What is the area of a triangle whose sides are 21 feet, 46 feet, and 50 feet long?

SOLUTION.—In order to apply the formula, suppose we let a represent the side that is 21 feet long; b, the side that is 50 feet long; and c, the side that is 46 feet long. Then substituting in the formula,

$$A = \frac{b}{2}\sqrt{a^2 - \left(\frac{a^2 + b^2 - c^2}{2b}\right)^2} = \frac{50}{2}\sqrt{21^2 - \left(\frac{21^2 + 50^2 - 46^2}{2 \times 50}\right)^2}$$

$$= \frac{50}{2}\sqrt{441 - \left(\frac{441 + 2{,}500 - 2{,}116}{100}\right)^2} = 25\sqrt{441 - \left(\frac{825}{100}\right)^2}$$

$$= 25\sqrt{441 - 8.25^2} = 25\sqrt{441 - 68.0625} = 25\sqrt{372.9375}$$

$$= 25 \times 19.312 = 482.8 \text{ square feet, nearly. Ans.}$$

The operations in the above examples have been extended much farther than was necessary; it was done in order to show the student every step of the process. The last formula is perfectly general, and the same answer would have been obtained had the 50-foot side been represented by a, the 46-foot side by b, and the 21-foot side by c.

3. The Rankine-Gordon formula for determining the least load in pounds that will cause a long column to break is

$$P = \frac{SA}{1 + q\dfrac{l^2}{G^2}},$$

in which $P =$ load (pressure) in pounds;

$S =$ ultimate strength (in pounds per square inch) of the material composing the column;

$A =$ area of cross-section of column in square inches;

$q =$ a factor (multiplier) whose value depends upon the shape of the ends of the column and on the material composing the column;

$l =$ length of column in inches;

and $G =$ least radius of gyration of cross-section of column.

The values of S, q, and G^2 are given in printed tables in books in which this formula occurs.

EXAMPLE.—What is the least load that will break a hollow wrought-iron column whose outside diameter is 14 inches; inside diameter, 11 inches; length, 20 feet, and whose ends are flat?

SOLUTION.—For steel, $S = 150{,}000$, and $q = \dfrac{1}{25{,}000}$ for flat-ended steel columns; A, the area of the cross-section, $= .7854\,(d_1^2 - d_2^2) = .7854\,(14^2 - 11^2)$, d_1 and d_2 being the outside and inside diameters,

respectively; $l = 20 \times 12 = 240$ inches; and $G^2 = \dfrac{d_1^2 + d_2^2}{16} = \dfrac{14^2 + 11^2}{16}$.
Substituting these values in the formula,

$$P = \dfrac{SA}{1 + q\dfrac{l^2}{G^2}} = \dfrac{150{,}000 \times .7854\,(14^2 - 11^2)}{1 + \dfrac{1}{25{,}000} \times \dfrac{240^2}{\dfrac{14^2 + 11^2}{16}}} =$$

$$\dfrac{150{,}000 \times 58.905}{1 + .1163} = \dfrac{8{,}835{,}750}{1.1163} = 7{,}915{,}211 \text{ pounds. Ans.}$$

4. EXAMPLE.—When $A = 10$, $B = 8$, $C = 5$, and $D = 4$, what is the value of E in the following:

(a) $E = \sqrt[3]{\dfrac{BCD}{A\left(2 + \dfrac{D^2}{C^2}\right)}}$?   (b) $E = \dfrac{A - \tfrac{3}{4}D + \dfrac{4B^2}{A + C}}{A - \sqrt{\dfrac{2B^2}{A + 22}}}$?

SOLUTION.—(a) Substituting,

$$E = \sqrt[3]{\dfrac{8 \times 5 \times 4}{10\left(2 + \dfrac{4^2}{5^2}\right)}}.$$

To simplify the denominator, square the 4 and 5, add the resulting fraction to 2, and multiply by 10. Simplifying, we have

$$E = \sqrt[3]{\dfrac{160}{10\left(2 + \dfrac{16}{25}\right)}} = \sqrt[3]{\dfrac{160}{10 \times \dfrac{66}{25}}} = \sqrt[3]{\dfrac{160}{\dfrac{660}{25}}} = \sqrt[3]{\dfrac{200}{33}}.$$

Reducing the fraction to a decimal before extracting the cube root,

$$E = \sqrt[3]{6.0606} = 1.823. \text{ Ans.}$$

(b) Substituting,

$$E = \dfrac{10 - \tfrac{3}{4} \times 4 + \dfrac{4 \times 8^2}{10 + 5}}{10 - \sqrt{\dfrac{2 \times 8^2}{10 + 22}}} = \dfrac{10 - 3 + \dfrac{4 \times 64}{15}}{10 - \sqrt{\dfrac{2 \times 64}{32}}} =$$

$$\dfrac{7 + 17.066 +}{10 - \sqrt{4}} = \dfrac{24.066 +}{8} = 3.008 +. \text{ Ans.}$$

345₁.—In the preceding pages, the unknown quantity has always been represented by the single letter at the left of the sign of equality, while the letters at the right have represented known values from which the required values could be found. It is possible, however, to find the value of the quantity represented by any letter in a formula, if the values represented by all the others are known. For example, let it be required to find how many strokes per minute an engine having a piston area of 78.54 sq. in. must make in order to develop 60 horsepower, if the mean effective pressure is 40 lb. per sq. in., and the length of stroke $1\frac{1}{4}$ ft. By substituting the given values in the formula $H = \dfrac{PLAN}{33,000}$, we have

$$60 = \dfrac{40 \times 1\frac{1}{4} \times 78.54 \times N}{33,000},$$

in which N, the number of strokes, is to be found.

But it is evident that the expression on the right of the sign of equality is equal to $\dfrac{40 \times 1\frac{1}{4} \times 78.54}{33,000} \times N$, a fraction whose numerator is composed of 3 factors. Reducing the numerator to a single number by performing the indicated multiplications, we obtain, after canceling,

$$60 = \dfrac{119}{1,000} \times N = .119\, N.$$

If 60 equals $.119\, N$, then N equals 60 divided by $.119$; hence,

$$N = \dfrac{60}{.119} = 504.2 \text{ revolutions per minute.}$$

The method of procedure is essentially the same when the unknown quantity occurs in the denominator of a formula. Thus, in the formula $f = \dfrac{m v^2}{r}$, suppose that $f = 375$, $m = 1.25$, and $v = 60$. Then, substituting,

$$375 = \dfrac{1.25 \times 60^2}{r} = \dfrac{4,500}{r}.$$

But if 375 equals 4,500 divided by r, then $375 \times r = 4,500$; hence, r must equal 4,500 divided by 375, or $r = \dfrac{4,500}{375} = 12$.

EXAMPLES FOR PRACTICE.

Find the numerical values of x in the following formulas, when $A = 9$, $B = 8$, $d = 10$, $e = 3$, and $c = 2$:

1. $x = \dfrac{d + c^2}{d^2 - 40}$.   Ans. $x = \frac{7}{30}$.

2. $x = \dfrac{\frac{1}{4}(A + e)}{c e}$.   Ans. $x = 1\frac{1}{4}$.

3. $x = \sqrt{\frac{d^2}{2c}} + \sqrt{AB^2}.$ Ans. $x = 29.$

4. $x = \frac{Ae}{\sqrt{16Bc}} + \frac{5}{16}.$ Ans. $x = 2.$

5. $x = (c + 2e)\left(\sqrt[3]{B} - \frac{1}{c}\right) + \frac{c^2 - c^2}{c^2 + c^2}.$ Ans. $x = 12\frac{5}{13}.$

6. $x = \sqrt{\frac{B c d}{A\left(2 + \frac{d^3}{c^2}\right)}}.$ Ans. $x = .396 +.$

MENSURATION.

346. Mensuration treats of the measurement of lines, angles, surfaces, and solids.

LINES AND ANGLES.

347. A **straight line** is one that does not change its direction throughout its whole length. To distinguish one straight line from another, its two extreme points are designated by letters. The line shown in Fig. 1 would be called the line AB.

348. A **curved line** changes its direction at every point. Curved lines are designated by three or more letters, as the curved line ABC, Fig. 2.

349. Parallel lines (Fig. 3) are those which are equally distant from each other at all points.

350. A line is **perpendicular** to another (see Fig. 4) when it meets that line so as not to incline towards it on either side.

351. A **vertical line** is one that points towards the center of the earth, and is also known as a *plumb* line.

352. A **horizontal line** (see Fig. 5) is one that makes a right angle with any vertical line.

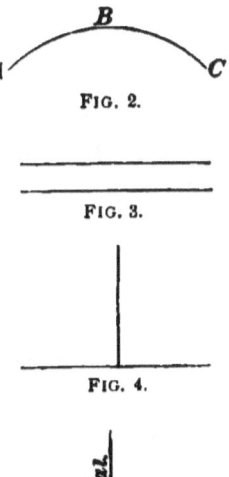

MENSURATION.

353. An **angle** is the opening between two lines which intersect or meet; the point of meeting is called the **vertex**

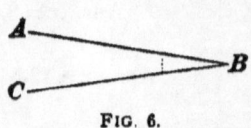

FIG. 6.

of the angle. Angles are distinguished by naming the vertex and a point on each line. Thus, in Fig. 6, the angle formed by the lines AB and CB is called the angle ABC, or the angle CBA; the letter at the vertex is always placed in the middle. When an angle stands alone so that it can not be mistaken for any other angle, only the vertex letter need be used. Thus, the angle referred to might be designated simply as the angle B.

354. If one straight line meets another straight line at a point between its ends, as in Fig. 7, two angles, ABC and ABD, are formed, which are called **adjacent angles**.

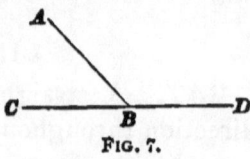

FIG. 7.

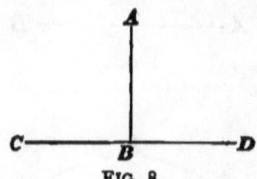

FIG. 8.

355. When these adjacent angles, ABC and ABD, are equal, as in Fig. 8, they are called **right angles**.

356. An **acute angle** is less than a right angle. ABC, Fig. 9, is an acute angle.

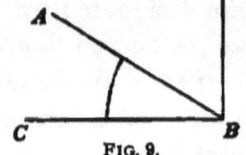

FIG. 9.

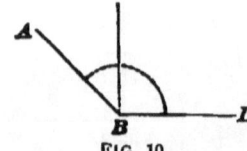

FIG. 10.

357. An **obtuse angle** is greater than a right angle. ABD (Fig. 10) is an obtuse angle.

358. A **circle** (see Fig. 11) is a figure bounded by a curved line, called the **circumference**, every point of which is equally distant from a point within, called the **center**.

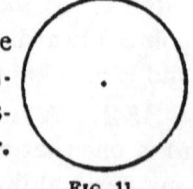

FIG. 11.

MENSURATION.

359. An **arc** of a circle is any part of its circumference; thus $a\,c\,b$, Fig. 12, is an arc of the circle.

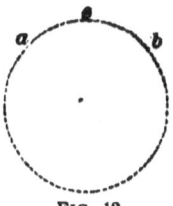

FIG. 12.

The circumference of every circle is considered to be divided into 360 equal parts, or arcs, called **degrees**; every degree is subdivided into 60 equal parts, called **minutes,** and every minute is again divided into 60 equal parts, called **seconds.**

Since 1 degree is $\frac{1}{360}$ of any circumference, it follows that the length of a degree will be different in circles of different sizes, but the proportion of the length of an arc of one degree to the whole circumference will always be the same, viz., $\frac{1}{360}$ of the circumference.

Degrees, minutes, and seconds are denoted by the symbols °, ′, ″. Thus, the expression 37° 14′ 44″ is read 37 degrees, 14 minutes, and 44 seconds.

360. The arcs of circles are used to measure angles. An angle having its vertex at the center of a circle is measured by the arc included between its sides; thus, in Fig. 13, the arc $F\,B$ measures the angle $F\,O\,B$. If the arc $F\,B$ contains 20°, or $\frac{20}{360}$ of the circumference, the angle $F\,O\,B$ would be an angle of 20°; if it contained 20° 14′ 18″, it would be an angle of 20° 14′ 18″, etc.

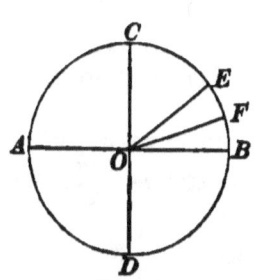

FIG. 13.

In the figure, if the line $C\,D$ be drawn perpendicular to $A\,B$, the adjacent angles will be equal, and the circle will be divided into four equal angles, each of which will be a right angle. A right angle, therefore, is an angle of $\frac{360°}{4}$, or 90°; two right angles are measured by 180°, or half the circumference, and four right angles by the whole circumference, or 360°. One-half of a right angle, as $E\,O\,B$, is an angle of 45°. An *acute* angle may now be defined as an angle of *less* than 90°, and an *obtuse* angle as one of *more* than 90°. These values are important, and should be remembered.

361. From the foregoing it will be evident that if a number of straight lines on the same side of a given straight line meet at the same point, the sum of all the angles formed is equal to two right angles, or 180°. Thus, in Fig. 14, angles $COB + DOC + EOD + FOE + AOF$ = 2 right angles, or 180°.

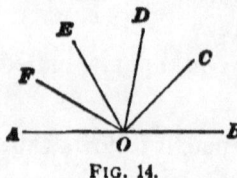

Fig. 14.

362. Also, if through a given point any number of straight lines be drawn, the sum of all the angles formed about the points of intersection equals four right angles, or 360°. Thus, in Fig. 15, angles $HOF + FOC + COA + AOG + GOE + EOD + DOB + BOH$ = four right angles, or 360°.

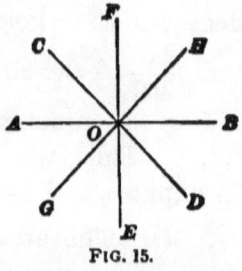

Fig. 15.

EXAMPLE.—In a fly-wheel, with 12 arms, how many degrees are there in the angle included between the center lines of any two arms, the arms being spaced equally?

SOLUTION.—Since there are 12 arms, there are 12 angles, which together equal 360°. Hence, one angle equals $\frac{1}{12}$ of 360°, or $\frac{360°}{12} = 30°$.

Ans.

EXAMPLES FOR PRACTICE.

1. How many seconds in 32° 14′ 6″? Ans. 116,046 sec.
2. How many degrees, minutes, and seconds do 38,582 seconds amount to? Ans. 10° 43′ 2″.
3. How many right angles are there in an angle of 170°? Ans. $1\frac{8}{9}$ right angles.
4. In a pulley with five arms, what part of a right angle is included between the center lines of any two arms? Ans. $\frac{4}{5}$ of a right angle.
5. If one straight line meets another so as to form an angle of 20° 10′, what does its adjacent angle equal? Ans. 159° 50′.
6. If a number of straight lines meet a given straight line at a given point, all being on the same side of the given line, so as to form six equal angles, how many degrees are there in each angle? Ans. 30°.

QUADRILATERALS.

363. A **plane figure** is any part of a plane or flat surface bounded by straight or curved lines.

364. A **quadrilateral** is a plane figure bounded by four straight lines.

365. A **parallelogram** is a quadrilateral whose opposite sides are parallel.

There are four kinds of parallelograms: the **square**, the **rectangle**, the **rhombus**, and the **rhomboid**.

366. A **rectangle** (Fig. 16) is a parallelogram whose angles are all right angles.

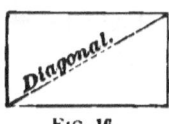

FIG. 16.

367. A **square** (Fig. 17) is a rectangle whose sides are all of the same length.

FIG. 17.

368. A **rhomboid** (Fig. 18) is a parallelogram whose opposite sides are equal and parallel, and whose angles are not right angles.

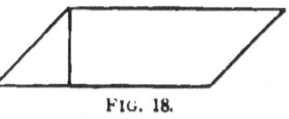

FIG. 18.

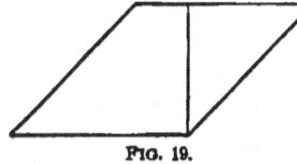

369. A **rhombus** (Fig. 19) is a parallelogram having equal sides, and whose angles are not right angles.

FIG. 19.

370. A **trapezoid** (Fig. 20) is a quadrilateral which has only two of its sides parallel.

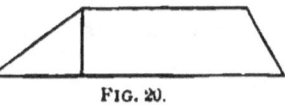

FIG. 20.

371. The **altitude** of a parallelogram is the perpendicular distance between the parallel sides, as shown by the dotted lines in Figs. 18, 19, and 20.

372. The **base** of *any* plane figure is the side on which it is supposed to stand.

MENSURATION.

373. The **area** of a surface is expressed by the number of unit squares it will contain.

374. A **unit square** is the square having a unit for its side. For example, if the unit is 1 inch, the unit square is the square each of whose sides measures 1 inch in length, and the area of a surface would be expressed by the number of square inches it would contain. If the unit were 1 foot, the unit square would measure 1 foot on each side, and the area of the given surface would be the number of square feet it would contain, etc.

The square that measures one inch on a side is called a **square inch,** and the one that measures one foot on a side is called a **square foot.** Square inch and square foot are abbreviated to sq. in. and sq. ft.

375. To find the area of any parallelogram:

Rule 44.—*Multiply the base by the altitude.*

NOTE.—Before multiplying, the base and altitude must be reduced to the same kind of units; that is, if the base should be given in feet and the altitude in inches, they could not be multiplied together until either the altitude had been reduced to feet, or the base to inches. This principle holds throughout the subject of mensuration.

EXAMPLE.—The sides of a square piece of sheet iron are each $10\frac{1}{4}$ inches long. How many square inches does it contain?

SOLUTION.—$10\frac{1}{4}$ inches $= 10.25$ inches when reduced to a decimal. The base and altitude are each 10.25 inches. Multiplying them together, $10.25 \times 10.25 = 105.06+$ sq. in. Ans.

EXAMPLE.—What is the area in square rods of a piece of land in the shape of a rhomboid, one side of which is 8 rods long, and whose length, measured on a line perpendicular to this side, is 200 feet?

SOLUTION.—The base is 8 rods and the altitude 200 feet. As the answer is to be in rods, the 200 feet should be reduced to rods. Reducing $200 \div 16\frac{1}{2} = 200 \div \frac{33}{2} = 12.12$ rods. Hence, area $= 8 \times 12.12 = 96.96$ sq. rd. Ans.

376. To find the area of a trapezoid:

Rule 45.—*Multiply one-half the sum of the parallel sides by the altitude.*

EXAMPLE.—A board 14 feet long is 20 inches wide at one end and 16 inches wide at the other. If the ends are parallel, how many square feet does the board contain?

MENSURATION.

SOLUTION.—One-half the sum of the parallel sides $= \frac{20+16}{2} = 18$ inches $= 1\frac{1}{2}$ feet. The length of the board corresponds to the altitude of a trapezoid. Hence, $14 \times 1\frac{1}{2} = 21$ sq. ft. Ans.

377. Having given the area of a parallelogram and one dimension, to find the other dimension:

Rule 46.—*Divide the area by the given dimension.*

EXAMPLE.—What is the width of a parallelogram whose area is 212 square feet and whose length is $26\frac{1}{2}$ feet?

SOLUTION.—$212 \div 26\frac{1}{2} = 212 \div \frac{53}{2} = 8$ feet. Ans.

The following examples illustrate a few special cases:

EXAMPLE.—An engine room is 22 feet by 32 feet. The engine-bed occupies a space of 3 feet by 12 feet; the fly-wheel pit, a space of 2 feet by 6 feet, and the outer bearing, a space of 2 feet by 4 feet. How many square feet of flooring will be required for the room?

SOLUTION.—Area of engine-bed $= 3 \times 12 = 36$ sq. ft.
Area of fly-wheel pit $= 2 \times 6 = 12$ sq. ft.
Area of outer bearing $= 2 \times 4 = \underline{8}$ sq. ft.
Total, $\overline{56}$ sq. ft.
Area of engine room $= 22 \times 32 = 704$ sq. ft.

$704 - 56 = 648$ square feet of flooring required. Ans.

EXAMPLE.—How many square yards of plaster will it take to cover the sides and ceiling of a room 16×20 feet and 11 feet high, having four windows, each 7×4 feet, and three doors, each 9×4 feet over all, the baseboard coming 6 inches above the floor?

SOLUTION.—
Area of ceiling $= 16 \times 20 = 320$ sq. ft.
Area of end walls $= 2(16 \times 11) = 352$ sq. ft.
Area of side walls $= 2(20 \times 11) = \underline{440}$ sq. ft.
Total area $= \overline{1,112}$ sq. ft.

From the above must be deducted:
Windows $= 4(7 \times 4) = 112$ sq. ft.
Doors $= 3(9 \times 4) = 108$ sq. ft.
Baseboard less the width of three doors $= (72-12) \times \frac{6}{12} = 30$ sq. ft.
Total number of feet to be deducted $= 112 + 108 + 30 = 250$ sq. ft.

Hence, number of square feet to be plastered $= 1,112 - 250 = 862$ sq. ft., or $95\frac{7}{9}$ sq. yd. Ans.

EXAMPLE.—How many acres are contained in a rectangular tract of land 800 rods long and 520 rods wide?

SOLUTION.—$800 \times 520 = 416,000$ sq. rd. Since there are 160 square rods in one acre, the number of acres $= 416,000 \div 160 = 2,600$ acres. Ans.

MENSURATION.

EXAMPLES FOR PRACTICE.

1. What is the area in square feet of a rhombus whose base is 84 inches, and whose altitude is 3 feet? Ans. 21 sq. ft.

2. A flat roof, 46 feet by 80 feet in size, is covered by tin roofing weighing one-half pound per square foot; what is the total weight of the roof? Ans. 1,840 lb.

3. One side of a room measures 16 ft. If the floor contains 240 square feet, what is the length of the other side? Ans. 15 ft.

4. How many square feet in a board 12 feet long, 18 inches wide at one end, and 12 inches wide at the other end? Ans. 15 sq. ft.

5. How much would it cost to lay a sidewalk a mile long and 8 feet 6 inches wide, at the rate of 20 cents per square foot? How much at the rate of $1.80 per square yard? Ans. $8,976 in each case.

6. How many square yards of plastering will be required for the ceiling and walls of a room 10 × 15 feet and 9 feet high; the room contains one door 3½ × 7 feet, three windows 3½ × 6 feet, and a baseboard 8 inches high? Ans. 53.5 sq. yd.

THE TRIANGLE.

378. A **triangle** is a plane figure having three sides.

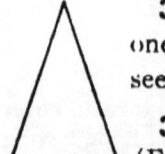
Fig. 21.

379. An **isosceles** triangle is one having two of its sides equal; see Fig. 21.

380. An **equilateral** triangle (Fig. 22) is one having all of its sides of the same length.

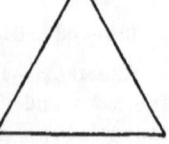
Fig. 22.

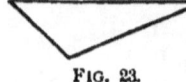
Fig. 23.

381. A **scalene** triangle (Fig. 23) is one having no two of its sides equal.

382. A **right-angled** triangle (Fig. 24) is any triangle having one right angle. The side opposite the right angle is called the **hypotenuse**.

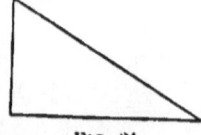
Fig. 24.

In any triangle the sum of the three angles equals two right angles, or 180°. Thus, in Fig. 25, the sum of the angles A, B, and C equals two right angles, or 180°. Hence, if any two angles of a triangle are given and it is required to find the third angle:

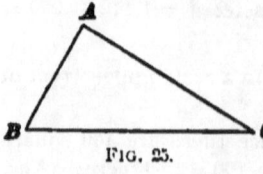
Fig. 25.

MENSURATION.

Rule 47.—*Add together the two given angles, and subtract their sum from 180°; the result will be the third angle.*

EXAMPLE.—If two angles of a triangle = 48° 16′ and 47° 50′ respectively, what does the third angle equal?

SOLUTION.—First reduce 48° 16′ and 47° 50′ to minutes, for convenience in adding and subtracting the angles. $48° = 48 \times 60 = 2,880′$; $2,880′ + 16′ = 2,896′$; hence, $48° 16′ = 2,896′$. In like manner, $47° 50′ = 47 \times 60 = 2,820′ + 50′ = 2,870′$. Adding the two angles together, and subtracting from 180° reduced to minutes, $2,896 + 2,870 = 5,766$; $180° = 180 \times 60 = 10,800′$; $10,800 - 5,766 = 5,034′$. Reducing this last number to degrees and minutes, $\frac{5,034}{60} = 83\frac{54}{60}° = 83° 54′$. Hence, the third angle in the triangle = 83° 54′.   Ans.

383. In any right-angled triangle there can be but one right angle, and since the sum of all the angles is two right angles, it is evident that the sum of the two acute angles must equal one right angle, or 90°. Therefore, if in any right-angled triangle one acute angle is known, to find the other acute angle:

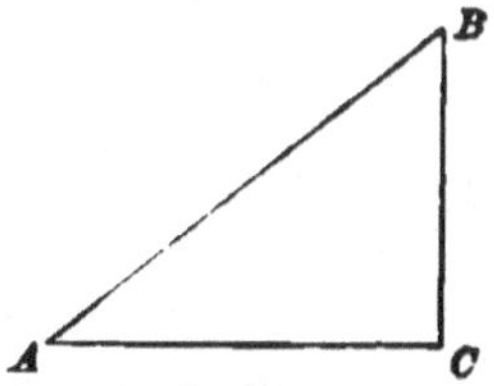

FIG. 26.

Rule 48.—*Subtract the known acute angle from 90°; the result will be the other acute angle.*

EXAMPLE.—If one acute angle, as A, of the right-angled triangle ABC, Fig. 26, equals 30°, what does the angle B equal?

SOLUTION.—$90° - 30° = 60°$.   Ans.

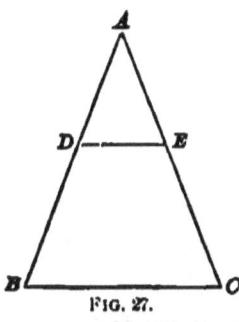

FIG. 27.

384. If a straight line be drawn through two sides of a triangle, parallel to the third side, a second triangle will be formed whose sides will be proportional to the corresponding sides of the first triangle. Thus, in the triangle ABC, Fig. 27, if the line DE be drawn parallel to the side BC, the triangle ADE will be formed and we shall have

(1) Side AD : side DE = side AB : side BC; and,

(2) Side AE : side DE = side AC : side BC; also,

(3) Side AD : side AE = side AB : side AC.

Example.—In Fig. 27, if $AB = 24$, $BC = 18$, and $DE = 8$, what does AD equal?

Solution.—Writing these values for the sides in (1),

$$AD : 8 = 24 : 18; \text{ whence } AD = \frac{24 \times 8}{18} = 10\tfrac{2}{3}. \text{ Ans.}$$

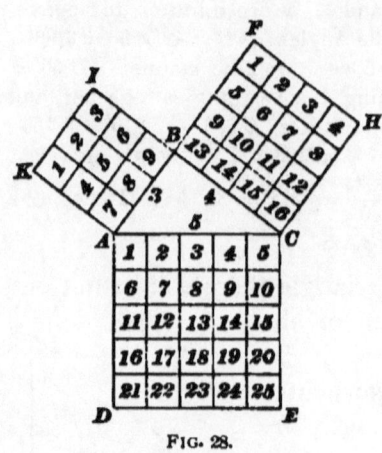

Fig. 28.

385. In any right-angled triangle, the square described on the hypotenuse is equal to the sum of the squares described upon the other two sides. If ABC, Fig. 28, is a right-angled triangle, right-angled at B, then the square described upon the hypotenuse AC is equal to the sum of the squares described upon the sides AB and BC. Hence, having given the two sides forming the right angle in a right-angled triangle, to find the hypotenuse:

Rule 49.—*Square each of the sides forming the right angle; add the squares together, and take the square root of the sum.*

Example.—If $AB = 3$ inches and $BC = 4$ inches, what is the length of the hypotenuse AC?

Solution.—Squaring each of the given sides, $3^2 = 9$ and $4^2 = 16$. Taking the square root of the sum of 9 and 16, the hypotenuse $= \sqrt{9+16} = \sqrt{25} = 5$ inches. Ans.

If the hypotenuse and one side are given, the other side can be found as follows:

Rule 50.—*Subtract the square of the given side from the square of the hypotenuse, and extract the square root of the remainder.*

Example.—The side given is 3 inches, the hypotenuse is 5 inches; what is the length of the other side?

Solution.—$3^2 = 9$; $5^2 = 25$. $25 - 9 = 16$, and the $\sqrt{16} = 4$ inches. Ans.

MENSURATION.

EXAMPLE.—If from a church steeple which is 150 feet high, a rope is to be attached to the top, and to a stake in the ground, which is 85 feet from the center of the base (the ground being supposed to be level), what must be the length of the rope?

SOLUTION.—In Fig. 29, AB represents the steeple, 150 feet high; C a stake 85 feet from the foot of the steeple, and AC the rope. Here we have a right-angled triangle, right-angled at B, and AC is the hypotenuse. The square of $AB = 150^2 = 22{,}500$; of CB, $85^2 = 7{,}225$. $22{,}500 + 7{,}225 = 29{,}725$; $\sqrt{29{,}725} = 172.4$ feet, nearly. Ans.

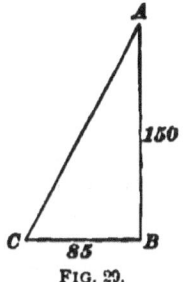

FIG. 29.

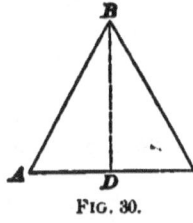

FIG. 30.

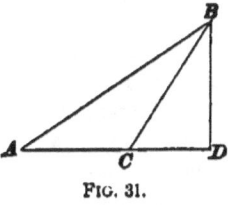

FIG. 31.

386. The **altitude** of any triangle is a line, as BD, drawn from the vertex B of the angle opposite the base AC, perpendicular to the base, as in Fig. 30, or to the base extended, as in Fig. 31.

If in any parallelogram a straight line, called the **diagonal**, be drawn, connecting two opposite corners it will divide the parallelogram into two equal triangles, as ADB and DBC in Fig. 32. The area of each triangle will equal one-half the area of the parallelogram, or one-half the product of the base and the altitude. Hence, to find the area of any triangle:

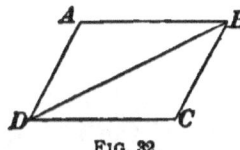

FIG. 32.

Rule 51.—*Multiply the base by the altitude, and divide the product by 2.*

EXAMPLE.—What is the area in square feet of a triangle whose base is 18 feet, and whose altitude is 7 feet 9 inches?

SOLUTION.—7 feet 9 inches $= 7\frac{3}{4}$ feet $= \frac{31}{4}$ feet. $18 \times \frac{31}{4} = 139\frac{1}{2}$, and one-half of $139\frac{1}{2} = 69\frac{3}{4}$ square feet. Ans.

To find the altitude or base of a triangle, having given the area and the base or altitude:

Rule 52.—*Multiply the area by 2, and divide by the given dimension.*

MENSURATION.

EXAMPLE.—What must be the height of a triangular piece of sheet metal to contain 100 square inches, if the base is 10 inches long?

SOLUTION.—$100 \times 2 = 200$; $200 \div 10 = 20$ inches. Ans.

EXAMPLES FOR PRACTICE.

1. What is the area of a triangle whose base is 18 feet long, and whose altitude is 10 feet 6 inches? Ans. 94.5 sq. ft.
2. Two angles of a scalene triangle together equal 100° 4'. What is the size of the third angle? Ans. 79° 56'.
3. One angle of a right-angled triangle equals 20° 10' 5". What is the size of the other acute angle? Ans. 69° 49' 55".
4. A ladder 65 feet long reaches to the top of a wall when its foot is 25 feet from the wall. How high is the wall? Ans. 60 ft.
5. Draw a triangle, and through two of its sides draw a line parallel to the base. Letter the different lines, and then, without referring to the text, write out the proportions existing between the sides of the two triangles.
6. A triangular piece of sheet metal weighs 24 pounds. If the base of the triangle is 4 feet and its height 6 feet, how much does the metal weigh per square foot? Ans. 2 lb.
7. The area of a triangle is 16 square inches. If the altitude is 4 inches, what does the base measure? Ans. 8 in.
8. Two sides of a right-angled triangle are 92 feet and 69 feet long. How long is the hypotenuse? Ans. 115 ft.

POLYGONS.

387. A **polygon** is a plane figure bounded by straight lines. The term is usually applied to a figure having more than four sides. The bounding lines are called the **sides**, and the sum of the lengths of all the sides is called the **perimeter** of the polygon.

388. A **regular polygon** is one in which all of the sides and all of the angles are equal.

389. A polygon of five sides is called a **pentagon**; one with six sides a **hexagon**; one of seven sides a **hep-**

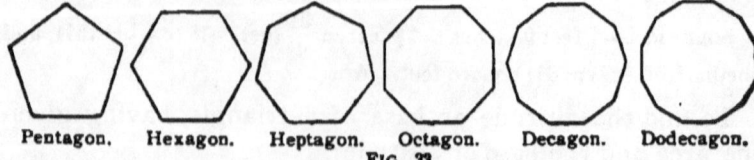

Pentagon. Hexagon. Heptagon. Octagon. Decagon. Dodecagon.
FIG. 33.

tagon, etc. Regular polygons having from five to twelve sides are shown in Fig. 33. In any polygon, the sum of all the interior angles, as $A + B + C + D + E$, Fig. 34, equals 180° multiplied by a number which is two less than the number of sides in the polygon. Hence, to find the size of any one of the interior angles of a *regular* polygon:

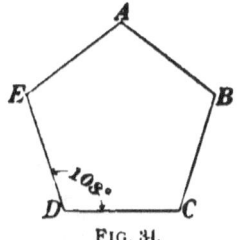

Fig. 34.

Rule 53.—*Multiply 180° by the number of sides less two, and divide the result by the number of sides; the quotient will be the number of degrees in each interior angle.*

EXAMPLE.—If Fig. 34 is a regular pentagon, how many degrees are there in each interior angle?

SOLUTION.—In a pentagon there are five sides; hence, $5 - 2 = 3$ and $180 \times 3 = 540$; $540 \div 5 = 108°$ in each angle. Ans.

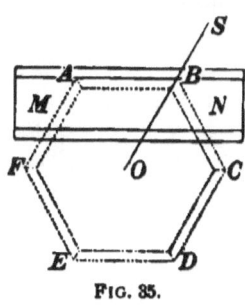

Fig. 35.

EXAMPLE.—It is desired to make a miter-box in which to cut a strip of molding to fit around a column having the shape of a regular hexagon. At what angle should the saw run across the miter-box?

SOLUTION.—In Fig. 35, let AB, BC, CD, etc., represent the pieces of molding as they will fit around the column. First find the size of one of the equal angles of the polygon by the above rule. Number of sides $= 6$; $6 - 2 = 4$; hence, $180 \times 4 = 720$, and $720 \div 6 = 120°$ in each angle. Now, let MN represent the miter-box, and OS the direction in which the saw should run; then, ABO is the angle made by the saw with the side of the miter-box; but, as the polygon is a regular one, this angle is one-half the interior angle ABC, which we have found to be 120°. Hence, the saw should run at an angle of $\frac{120}{2} = 60°$ with the side of the miter-box.

390. The area of any regular polygon may be found by drawing lines from the center to each angle, and computing the area of each triangle thus formed. Hence, to find the area of any regular polygon:

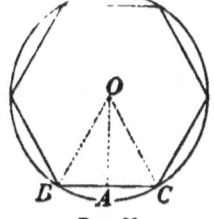

Fig. 36.

Rule 54.—*Multiply the length of a side by half the distance from the side to the center, and that product by the number of sides. The last product will be the area of the figure.*

EXAMPLE.—In Fig. 36 the side BC of the regular hexagon is 12 inches and the distance AO is 10.4 inches; required, the area of the polygon.

SOLUTION.—$10.4 \div 2 = 5.2$; $12 \times 5.2 \times 6 = 374.4$ sq. in. Ans.

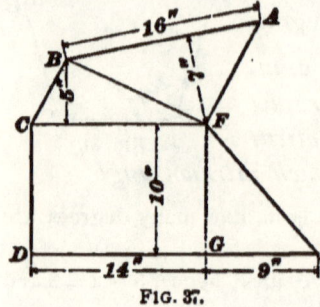

FIG. 37.

391. To obtain the area of any irregular polygon, draw diagonals dividing the polygon into triangles and quadrilaterals, and compute the areas of these separately; their sum will be the area of the figure.

EXAMPLE.—It is required to find the area of the polygon $ABCDEF$, Fig. 37.

SOLUTION.—Draw the diagonals BF and CF and the line FG perpendicular to DE, dividing the figure into the triangles ABF, BCF, and FGE, and the rectangle $FCDG$. Let it be supposed that the altitudes of the figures and the lengths of the sides AB, DG, and GE are as indicated in the polygon above. Then,

$$\text{Area } ABF = \frac{16 \times 7}{2} = 56 \text{ sq. in.}$$

$$\text{Area } BCF = \frac{14 \times 5}{2} = 35 \text{ sq. in.}$$

$$\text{Area } FCDG = 14 \times 10 = 140 \text{ sq. in.}$$

$$\text{Area } FGE = \frac{9 \times 10}{2} = 45 \text{ sq. in.}$$

Total area $= 56 + 35 + 140 + 45 = 276$ sq. in. Ans.

EXAMPLES FOR PRACTICE.

1. How many degrees are there in one of the angles of a regular octagon? Ans. 135°.

2. Find the area of the polygon $ABCDEF$ (see Fig. 37), supposing each of the given dimensions to be increased to 1¼ times the length given in the figure. Ans. 621 sq. in.

3. What is the area of a regular heptagon, whose sides are 4 inches long, the distance from one side to the center being 4.15 inches?
Ans. 58.1 sq. in.

4. At what angle should the saw run in a miter-box to cut strips to fit around the edge of a table top, made in the shape of a regular pentagon? Ans. 54°.

MENSURATION.

THE CIRCLE.

392. A **circle** (Fig. 38) is a figure bounded by a curved line, called the **circumference,** every point of which is equally distant from a point within, called the **center.**

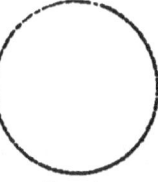
FIG. 38.

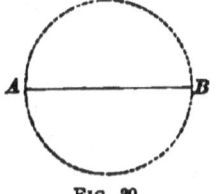
FIG. 39.

393. The **diameter** of a circle is a straight line passing through the center and terminated at both ends by the circumference; thus, AB (Fig. 39) is a diameter of the circle.

394. The **radius** of a circle, AO (Fig. 40), is a straight line drawn from the center O to the circumference. It is equal in length to one-half the diameter. The plural of radius is **radii,** and all radii of a circle are equal.

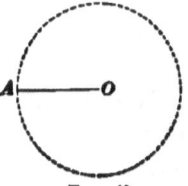
FIG. 40.

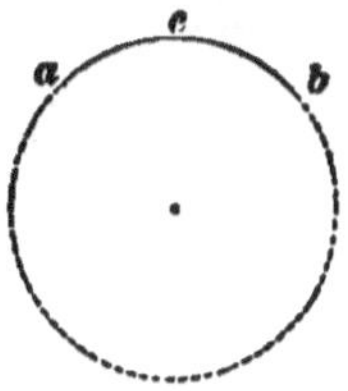

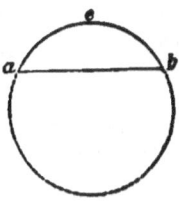
FIG. 41.

395. An **arc** of a circle (see $a\,e\,b$, Fig. 41) is any part of its circumference.

396. A **chord** is a straight line joining any two points in a circumference; or it is a straight line joining the extremities of an arc; thus, the straight line ab, Fig. 42, is a chord of the circle whose corresponding arc is $a\,e\,b$.

FIG. 42.

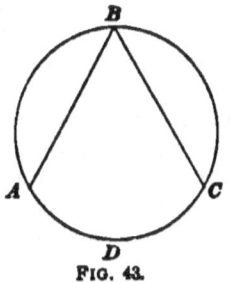
FIG. 43.

397. An **inscribed angle** is one whose vertex lies on the circumference of a circle, and whose sides are chords. It is measured by one-half the intercepted arc. Thus, in Fig. 43, ABC is an inscribed angle, and it is measured by one-half the arc ADC.

EXAMPLE.—If in Fig. 43, the arc $ADC = \frac{2}{5}$ of the circumference, what is the measurement of the inscribed angle ABC?

SOLUTION.—Since the angle is an inscribed angle, it is measured by one-half the intercepted arc, or $\frac{2}{5} \times \frac{1}{2} = \frac{1}{5}$ of the circumference. The whole circumference $= 360°$; hence, $360° \times \frac{1}{5} = 72°$; therefore, angle ABC is an angle of $72°$.

398. If a circle is divided into halves, each half is called a **semicircle**, and each half circumference is called a **semi-circumference**.

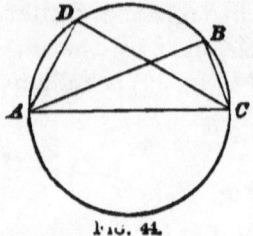

FIG. 44.

Any angle inscribed in a semicircle is a right angle, since it is measured by one-half a semi-circumference, or $180° \div 2 = 90°$. Thus, the angles ADC and ABC, Fig. 44, are right angles, since they are inscribed in a semicircle.

399. An **inscribed** polygon is one whose vertexes lie on the circumference of a circle, and whose sides are chords, as $ABCDE$, Fig. 45.

The sides of an inscribed regular hexagon have the same length as the radius of the circle.

If, in any circle, a radius be drawn perpendicular to any chord, it bisects (cuts in halves) the chord. Thus, if the radius OC, Fig. 46, is perpendicular to the chord AB, $AD = DB$.

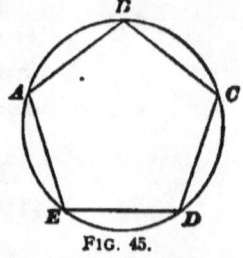

FIG. 45.

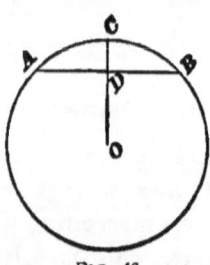

FIG. 46.

EXAMPLE.—If a regular pentagon be inscribed in a circle, and a radius is drawn perpendicular to one of the sides, what are the lengths of the two parts of the side, the perimeter of the pentagon being 27 inches?

SOLUTION.—A pentagon has five sides, and since it is a regular pentagon, all the sides are of equal lengths; the perimeter of a pentagon, which equals the distance around it, or equals the sum of all the sides, is 27 inches. Therefore, the length of one side $= 27 \div 5 = 5\frac{2}{5}$ inches. Since the pentagon is an inscribed pentagon, its sides are chords, and as a radius perpendicular to a chord bisects it, we have $5\frac{2}{5} \div 2 = 2\frac{7}{10}$ inches, which equals the length of each of the parts of the side, cut by a radius perpendicular to it.

MENSURATION.

400. If, from any point on the circumference of a circle, a perpendicular be let fall upon a given diameter, it will divide the diameter into two parts, one of which will be in the same ratio to the perpendicular as the perpendicular is to the other. That is, the perpendicular will be a *mean proportional* between the two parts.

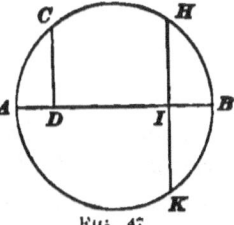

FIG. 47.

If AB, Fig. 47, is the given diameter, and C any point on the circumference, then $AD:CD::CD:DB$, CD being a mean proportional between AD and DB.

EXAMPLE.—If $HK = 30$ feet, and $IB = 8$ feet, what is the diameter of the circle, HK being perpendicular to AB?

SOLUTION.—30 feet ÷ 2 feet = 15 feet = IH. And $BI:IH::IH:IA$, or $8:15::15:IA$.

Therefore, $IA = \dfrac{15^2}{8} = \dfrac{225}{8} = 28\frac{1}{8}$ feet, and $IA + IB = 28\frac{1}{8} + 8 = 36\frac{1}{8}$ feet $= AB$ the diameter of the circle. Ans.

401. When the diameter of a circle and the lengths of the two parts into which it is divided are given, the length of the perpendicular may be found by multiplying the lengths of the two parts together and extracting the square root of the product.

EXAMPLE.—In Fig. 47, the diameter of the circle AB is $36\frac{1}{8}$ feet, and the distance BI is 8 feet; what is the length of the line HK?

SOLUTION.—As the diameter of the circle is $36\frac{1}{8}$ feet, and as BI is 8 feet, IA is equal to $36\frac{1}{8} - 8 = 28\frac{1}{8}$ feet. The two parts, therefore, are 8 and $28\frac{1}{8}$ feet, and their product $= 8 \times 28\frac{1}{8} = 8 \times \dfrac{225}{8} = 225$; the square root of their product $= \sqrt{225} = 15$ feet, and as $HK = IH + IK$, or $2\,IH$, $HK = 15 \times 2 = 30$ feet. Ans.

402. To find the circumference of a circle, the diameter being given:

Rule 55.—*Multiply the diameter by 3.1416.*

EXAMPLE.—What is the circumference of a circle whose diameter is 15 inches?

SOLUTION.—$15 \times 3.1416 = 47.124$ inches. Ans.

403. To find the diameter of a circle, the circumference being given:

Rule 56.—*Divide the circumference by 3.1416.*

EXAMPLE.—What is the diameter of a circle whose circumference is 65.973 inches?

SOLUTION.—65.973 ÷ 3.1416 = 21 inches. Ans.

404. To find the length of an arc of a circle:

Rule 57.—*Multiply the length of the circumference of the circle of which the arc is a part by the number of degrees in the arc, and divide by 360.*

EXAMPLE.—What is the length of an arc of 24°, the radius of the arc being 18 in.?

SOLUTION.—18 × 2 = 36 in. = the diameter of the circle. 36 × 3.1416 = 113.1 in., the circumference of the circle of which the arc is a part.

113.1 × $\frac{24}{360}$ = 7.54 in., or the length of the arc. Ans.

405. To find the area of a circle:

Rule 58.—*Square the diameter, and multiply by .7854.*

EXAMPLE.—What is the area of a circle whose diameter is 15 inches?
SOLUTION.—15^2 = 225; and 225 × .7854 = 176.72 sq. in. Ans.

406. Given the area of a circle, to find its diameter:

Rule 59.—*Divide the area by .7854, and extract the square root of the quotient.*

EXAMPLE.—The area of a circle = 17,671.5 square inches. What is its diameter in feet?

SOLUTION.—$\sqrt{\frac{17,671.5}{.7854}}$ = 150 inches.

$\frac{150}{12}$ = 12¼ feet, or the diameter. Ans.

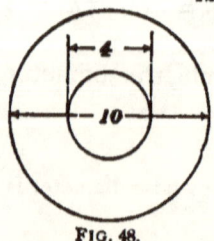

FIG. 48.

EXAMPLE.—What is the area of a flat circular ring, Fig. 48, whose outside diameter is 10 inches, and whose inside diameter is 4 inches?

SOLUTION.—The area of the large circle = 10^2 × .7854 = 78.54 square inches; the area of the small circle = 4^2 × .7854 = 12.57 square inches. The area of the ring is the difference between these areas, or 78.54 − 12.57 = 65.97 square inches. Ans.

407. To find the area of a sector (a **sector** of a circle is the area included between two radii and the circumference, as, for example, the area $C\ O\ E$, Fig. 13):

MENSURATION.

Rule 60.—*Divide the number of degrees in the arc of the sector by 360. Multiply the result by the area of the circle of which the sector is a part.*

EXAMPLE.—The number of degrees in the angle formed by drawing radii from the center of a circle to the extremities of the arc of the circle is 75°. The diameter of the circle is 12 inches; what is the area of the sector?

SOLUTION.—$\frac{75}{360} = \frac{5}{24}$; and $12^2 \times .7854 = 113.1$ square inches.

$113.1 \times \frac{5}{24} = 23.56$ square inches, the area. Ans.

408. To find the area of a segment of a circle (a **segment** of a circle is the area included between a chord and its arc; for example, the area $A\,B\,C$, Fig. 49):

Rule 61.—*Divide the diameter by the height of the segment; subtract .608 from the quotient, and extract the square root of the remainder. This result multiplied by 4 times the square of the height of the segment, and then divided by 3, will give the area, very nearly.*

The rule, expressed as a formula, is as follows, where $D =$ the diameter of the circle, and $h =$ the height of the segment, see Fig. 49:

Area of $A\,B\,C\,A = \frac{4h^2}{3} \sqrt{\frac{D}{h} - .608}$.

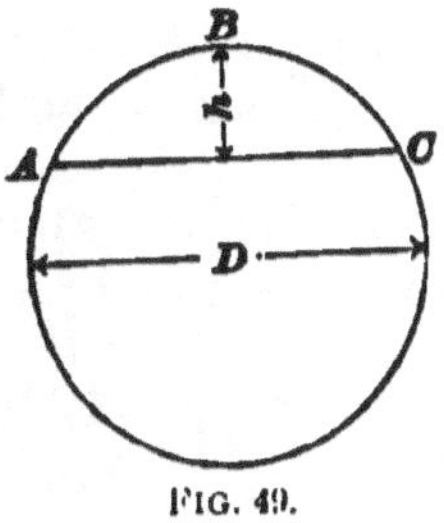

EXAMPLE.—What is the area of the segment of a circle whose diameter is 54 inches, the height of the segment being 20 inches?

SOLUTION.—Substituting in the formula,

Area $= \frac{4 \times 20^2}{3} \sqrt{\frac{54}{20} - .608}$, $\frac{4 \times 20^2}{3} = \frac{4 \times 400}{3} =$

FIG. 49.

$\frac{1,600}{3}$; $\sqrt{\frac{54}{20} - .608} = \sqrt{2.092} = 1.447$; $\frac{1,600}{3} \times 1.447 = 771.7$ sq. in.

Ans.

EXAMPLES FOR PRACTICE.

1. An angle inscribed in a circle intercepts one-third of the circumference. How many degrees are there in the angle? Ans. 60°.

2. Suppose that in Fig. 47, the diameter $A\,B = 15$ feet, and the distance $B\,I = 3$ feet. What is the length of the line $H\,K$?

Ans. 12 ft.

138 MENSURATION.

3. The diameter of a fly-wheel is 18 feet. What is the distance around it to the nearest 16th of an inch? Ans. 56 ft. 6$\frac{9}{16}$ in.

4. A carriage wheel was observed to make 71$\frac{2}{3}$ turns while going 300 yards. What was its diameter? Ans. 4 ft., nearly.

5. What is the length of an arc of 64°, the radius of the arc being 30 inches? Ans. 33.51 in.

6. Find the area of a circle 2 feet 3 inches in diameter.
Ans. 3.976 sq. ft.

7. What must be the diameter of a circle to contain 100 square inches? Ans. 11.28 in.

8. Compute the area of a segment, whose height is 11 inches, and the radius of whose arc is 21 inches. Ans. 289.04 sq. in.

9. Find the area of a flat circular ring whose outside diameter is 12 inches and whose inside diameter is 6 inches. Ans. 84.82 sq. in.

THE PRISM AND CYLINDER.

409. A **solid**, or body, has three dimensions: length, breadth, and thickness. The sides which enclose it are called the **faces**, and their intersections are called **edges**.

410. A **prism** is a solid whose ends are equal and parallel polygons, and whose sides are parallelograms. Prisms take their names from the form of their bases. Thus, a triangular prism is one having a triangle for its base; a hexagonal prism is one having a hexagon for its base, etc.

411. A **cylinder** is a body of uniform diameter whose ends are equal parallel circles.

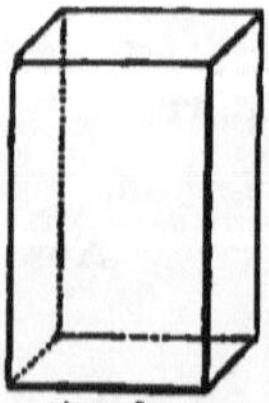

412. A **parallelopipedon** (Fig. 50) is a prism whose bases (ends) are parallelograms.

413. A **cube** (Fig. 51) is a prism whose faces and ends are squares. All the faces of a cube are equal.

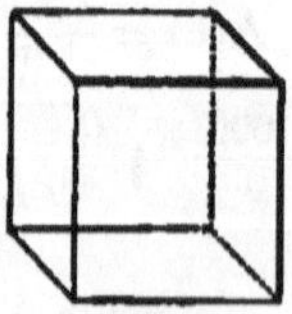

FIG. 50. FIG. 51.

In the case of plane figures, we have had to do with perimeters and areas. In the case of solids, we have to do with the areas of their outside surfaces, and with their contents or volumes.

MENSURATION.

414. The **entire surface** of any solid is the area of the whole outside of the solid, including the ends.

The **convex surface** of a solid is the same as the entire surface, except that the areas of the ends are not included.

415. A **unit of volume** is a cube each of whose edges is equal in length to the unit. The **volume** is expressed by the number of times it will contain a *unit of volume*.

Thus, if the unit of length is 1 inch, the unit of volume will be the cube whose edges each measure 1 inch, this cube being 1 *cubic inch;* and the number of cubic inches the solid contains will be its volume. If the unit of length is 1 foot, the unit of volume will be 1 *cubic foot*, etc. Cubic inch, cubic foot, and cubic yard are abbreviated to cu. in., cu. ft., and cu. yd., respectively.

Instead of the word *volume*, the expression **cubical contents** is sometimes used.

416. To find the area of the convex surface of a prism or cylinder:

Rule 62.—*Multiply the perimeter of the base by the altitude.*

EXAMPLE.—A block of marble is 24 inches long, and its ends are 9 inches square. What is the area of its convex surface?

SOLUTION.—$9 \times 4 = 36 =$ the perimeter of the base; $36 \times 24 = 864$ sq. in., the convex area. Ans.

To find the entire area of the outside surface, add the areas of the two ends to the convex area. Thus, the area of the two ends $= 9 \times 9 \times 2 = 162$ sq. in.; $864 + 162 = 1,026$ sq. in. Ans.

EXAMPLE.—How many square feet of sheet iron will be required for a pipe 1½ feet in diameter and 10 feet long, neglecting the amount necessary for lapping?

SOLUTION.—The problem is to find the convex surface of a cylinder 1½ feet in diameter and 10 feet long. The perimeter, or circumference, of the base $= 1\frac{1}{2} \times 3.1416 = 1.5 \times 3.1416 = 4.712$ feet. The convex surface $= 4.712 \times 10 = 47.12$ sq. ft. of metal. Ans.

417. To find the volume of a prism or a cylinder:

Rule 63.—*Multiply the area of the base by the altitude.*

EXAMPLE.—What is the weight of a length of wrought-iron shafting 16 feet long and 2 inches in diameter? Wrought iron weighs .28 pound per cubic inch.

MENSURATION.

Solution.—The shaft is a cylinder 16 feet long. The area of one end, or the base, $= 2^2 \times .7854 = 3.1416$ sq. in. Since the weight of the iron is given per cubic inch, the contents of the shaft must be found in cubic inches. The length, 16 feet, reduced to inches $= 16 \times 12 = 192$ in.; $3.1416 \times 192 = 603.19$ cu. in. $=$ the volume. The weight $= 603.19 \times .28 = 168.89$ lb. Ans.

Example.—Find the cubical contents of a hexagonal prism, Fig. 52, 12 inches long, each edge of the base being one inch long.

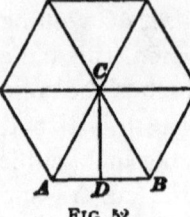

Fig. 52.

Solution.—In order to obtain the area of one end, the distance CD from the center C to one side must be found.

In the right-angled triangle CDA, side $AD = \frac{1}{2}AB$, or one-half inch, and since the polygon is a hexagon, side $CA =$ distance AB, or one inch. Hence, CA being the hypotenuse, the length of side $CD = \sqrt{1^2 - (\frac{1}{2})^2} = \sqrt{1^2 - .5^2} = \sqrt{.75}$, or .866 inch. Area of triangle $ACB = \dfrac{1 \times .866}{2} = .433$ sq. in.; area of the whole polygon $= .433 \times 6 = 2.598$ sq. in. Hence, the contents of the prism $= 2.598 \times 12 = 31.176$ cu. in. Ans.

Example.—It is required to find the number of cubic feet of steam space in the boiler shown in Fig. 53. The boiler is 16 feet long between

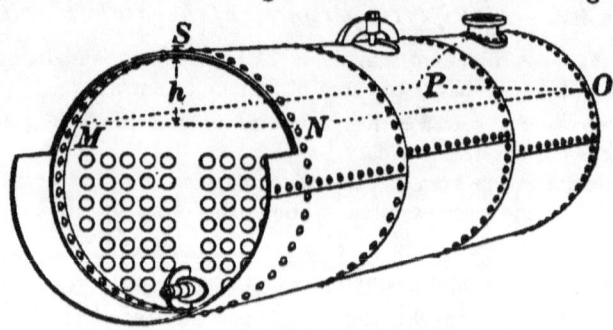

Fig. 53.

heads, 54 inches in diameter, and the mean water line MN is at a distance of 16 inches from the top of the boiler. The volume of the steam outlet casting may be neglected.

Solution.—The volume of the steam space, which is that space within the boiler above the surface $MNOP$ of the water, is found by the rule for finding the volume of a prism or cylinder, the area MNS being the base, and the length NO the altitude. First obtain the area of the segment MNS, whose height h is 16 inches, in square feet; then multiply the result by 16, the length of the boiler.

By the formula previously given, the area of the segment $=$

MENSURATION. 141

$$\frac{4h^2}{3}\sqrt{\frac{D}{h} - .608} = \frac{4 \times 16^2}{3}\sqrt{\frac{54}{16} - .608};$$

$$\frac{4 \times 16^2}{3} = 341.33;\ \sqrt{\frac{54}{16} - .608} = \sqrt{2.767} = 1.663.$$

Hence, the area = 341.33 × 1.663 = 567.63 sq. in. This, reduced to square feet, = 567.63 ÷ 144 = 3.942 sq. ft., and the volume, therefore, = 3.942 × 16 = 63.07 cu. ft. Ans.

In the above solution, the space occupied by the stays is not considered, for sake of simplicity. They are not shown in the figure.

EXAMPLE.—In the above boiler there are 60 tubes, 3¼ inches outside diameter. How many gallons of water will it take to fill the boiler up to the mean water level, there being 231 cubic inches in a gallon?

SOLUTION.—Find the volume in cubic inches of that part of the boiler below the surface of the water $MNOP$, since the contents of a gallon is given in cubic inches, and from it subtract the volume of the tubes in cubic inches.

This may be done by first finding the *total* area of one end of the boiler in square inches, from it subtracting the area of the segment MNS, and the areas of the ends of the tubes in square inches, and then by multiplying the result by the length of the boiler *in inches*.

Total area of one end = $54^2 \times .7854 = 2,290.23$ sq. in.
Area of segment MNS, as found in last example, = 567.63 sq. in.
Area of the end of one tube = $3.25^2 \times .7854 = 8.2958$ sq. in.
Area of the ends of the 60 tubes = $8.2958 \times 60 = 497.75$ sq. in.

Hence, the area to be subtracted = 567.63 + 497.75 = 1,065.38 sq. in. Subtracting, 2,290.23 − 1,065.38 = 1,224.85 sq. in. = net area.

The cubical contents = 1,224.85 × 16 × 12 = 235,171.2 cu. in. This, divided by 231, will give the number of gallons; whence, 235,171.2 ÷ 231 = 1,018.06 gallons of water. Ans.

EXAMPLES FOR PRACTICE.

1. Find the area in square inches of the convex surface of a bar of iron 4¼ inches in diameter, and 8 feet 5 inches long. Ans. 1,348.53 sq. in.

2. Find the area of the entire surface of the above bar.
Ans. 1,376.9 sq. in.

3. What is the area of the entire surface of the hexagonal prism whose base is shown in Fig. 52? Ans. 77.196 sq. in.

4. A multitubular boiler has the following dimensions: diameter, 50 inches; length between heads, 15 feet; number of tubes, 56; outside diameter of tubes, 3 inches; distance of mean water line from top of boiler, 10 inches. (*a*) Compute the steam space in cubic feet. (*b*) Find the number of gallons of water required to fill the boiler up to the mean water line. Ans. { (*a*) 56.4 cu. ft.
{ (*b*) 800 gallons.

THE PYRAMID AND CONE.

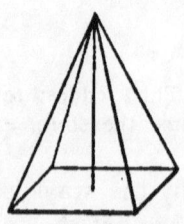

FIG. 54.

418. A **pyramid** (Fig. 54) is a solid whose base is a polygon, and whose sides are triangles uniting at a common point, called the **vertex**.

419. A **cone** (Fig. 55) is a solid whose base is a circle and whose convex surface tapers uniformly to a point called the **vertex**.

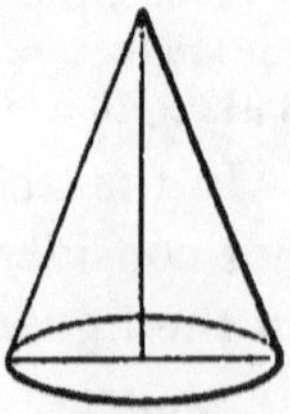

FIG. 55.

420. The **altitude** of a pyramid or cone is the perpendicular distance from the vertex to the base.

421. The **slant height** of a *pyramid* is a line drawn from the vertex perpendicular to one of the sides of the base. The slant height of a *cone* is any straight line drawn from the vertex to the circumference of the base.

422. To find the convex area of a pyramid or cone:

Rule 64.—*Multiply the perimeter of the base by one-half the slant height.*

EXAMPLE.—What is the convex area of a pentagonal pyramid, if one side of the base measures 6 inches, and the slant height = 14 inches?

SOLUTION.—The base of a pentagonal pyramid is a pentagon, and, consequently, has five sides.

$6 \times 5 = 30$ inches, or the perimeter of the base. $30 \times \frac{14}{2} = 210$ sq. in., or the convex area. Ans.

EXAMPLE.—What is the entire area of a right cone whose slant height is 17 inches, and whose base is 8 inches in diameter?

SOLUTION.—The perimeter of the base $= 8 \times 3.1416 = 25.1328$ in.

$$\text{Convex area} = 25.1328 \times \frac{17}{2} = 213.63 \text{ sq. in.}$$
$$\text{Area of base} = 8^2 \times .7854 = 50.27 \text{ sq. in.}$$
$$\text{Entire area} = 263.90 \text{ sq. in.} \quad \text{Ans.}$$

423. To find the volume of a pyramid or cone:

Rule 65.—*Multiply the area of the base by one-third of the altitude.*

EXAMPLE.—What is the volume of a triangular pyramid, each edge of whose base measures 6 inches, and whose altitude is 8 inches?

MENSURATION. 143

SOLUTION.—Draw the base as shown in Fig. 56; it will be an equilateral triangle, all of whose sides are 6 inches long.

Draw a perpendicular BD from the vertex to the base; it will divide the base into two equal parts, since an equilateral triangle is also isosceles, and will be the altitude of the triangle. In order to obtain the area of the base, this altitude must be determined.

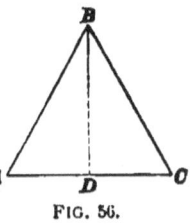

FIG. 56.

In the right-angled triangle BDA, the hypotenuse $BA = 6$ inches, and side $AD = 3$ inches, to find the other side,

$$BD = \sqrt{6^2 - 3^2} = 5.2 \text{ inches, nearly.}$$

Area of the base, or $BAC, = \dfrac{6 \times 5.2}{2} = 15.6$ sq. in. Hence, the volume $= 15.6 \times \dfrac{8}{3} = 41.6$ cu. in. Ans.

EXAMPLE.—What is the volume of a cone whose altitude is 18 inches, and whose base is 14 inches in diameter?

SOLUTION.—Area of the base $= 14^2 \times .7854 = 153.94$ sq. in. Hence, the volume $= 153.94 \times \dfrac{18}{3} = 923.64$ cu. in. Ans.

EXAMPLES FOR PRACTICE.

1. Find the convex surface of a square pyramid whose slant height is 28 inches, and one edge of whose base is $7\frac{1}{2}$ inches long.

Ans. 420 sq. in.

2. What is the volume of a triangular pyramid, one edge of whose base measures 3 inches, and whose altitude is 4 inches?

Ans. 5.2 cu. in.

3. Find the volume of a cone whose altitude is 12 inches, and the circumference of whose base is 31.416 inches. Ans. 314.16 cu. in.

NOTE.—Find the diameter of the base and then its area.

THE FRUSTUM OF A PYRAMID OR CONE.

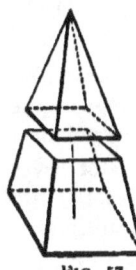

FIG. 57.

424. If a pyramid be cut by a plane, parallel to the base, so as to form two parts, as in Fig. 57, the lower part is called the **frustum** of the pyramid.

If a cone be cut in a similar manner, as in Fig. 58, the lower part is called the **frustum** of the cone.

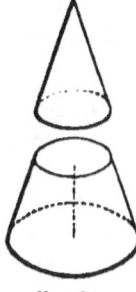

FIG. 58.

425. The upper end of the frustum of a pyramid or cone is called the **upper base,** and the lower end the **lower base.** The **altitude** of a frustum is the perpendicular distance between the bases.

426. To find the convex surface of a frustum of a pyramid or cone:

Rule 66.—*Multiply one-half the sum of the perimeters of the two bases by the slant height of the frustum.*

EXAMPLE.—Given, the frustum of a triangular pyramid, in which one side of the lower base measures 10 inches, one side of the upper base measures 6 inches, and whose slant height is 9 inches; find the area of the convex surface.

SOLUTION.—10 in. $\times$ 3 = 30 in., the perimeter of the lower base.
6 in. $\times$ 3 = 18 in., the perimeter of the upper base.
$\frac{30 + 18}{2} = 24$ in., or one-half the sum of the perimeters of the two bases. $24 \times 9 = 216$ sq. in., the convex area. Ans.

EXAMPLE.—If the diameters of the two bases of a frustum of a cone are 12 inches and 8 inches, respectively, and the slant height is 12 inches, what is the entire area of the frustum?

SOLUTION.—$\frac{(12 \times 3.1416) + (8 \times 3.1416)}{2} \times 12 = 376.99$ sq. in., the area of the convex surface.

Area of the upper base = $8^2 \times .7854 = 50.27$ sq. in.
Area of the lower base = $12^2 \times .7854 = 113.1$ sq. in.
The entire area of the frustum = $376.99 + 50.27 + 113.1 = 540.36$ sq. in. Ans.

427. To find the volume of the frustum of a pyramid or cone:

Rule 67.—*Add together the areas of the upper and lower bases, and the square root of the product of the two areas; multiply the sum by one-third of the altitude.*

EXAMPLE.—Given, a frustum of a square pyramid (one whose base is a square); each edge of the lower base is 12 inches, each edge of the upper base is 5 inches, and its altitude is 16 inches; what is its volume?

SOLUTION.—Area of upper base = $5 \times 5 = 25$ sq. in.; area of lower base = $12 \times 12 = 144$ sq. in.; the square root of the product of the two areas = $\sqrt{25 \times 144} = 60$. Adding these three results, and multiplying by one-third the altitude, $25 + 144 + 60 = 229$; $229 \times \frac{16}{3} = 1,221\frac{1}{3}$ cu. in. = the volume. Ans.

MENSURATION.

EXAMPLE.—How many gallons of water will a round tank hold, which is 4 feet in diameter at the top, 5 feet in diameter at the bottom, and 8 feet deep?

SOLUTION.—There are 231 cubic inches in a gallon, and the volume of the tank should be found in cubic inches. The tank is in the shape of the frustum of a cone. The upper diameter $= 4 \times 12 = 48$ inches; the lower diameter $= 5 \times 12 = 60$ inches, and the depth $= 8 \times 12 = 96$ inches. Area of upper base $= 48^2 \times .7854 = 1,809.56$ sq. in.; area of lower base $= 60^2 \times .7854 = 2,827.44$ sq. in.; $\sqrt{1,809.56 \times 2,827.44} = 2,261.95$.

Whence, $1,809.56 + 2,827.44 + 2,261.95 = 6,898.95$; $6,898.95 \times \dfrac{96}{3} = 220,766.4$ cu. in. $=$ contents. Now, since there are 231 cu. in. in one gallon, the tank will hold $220,766.4 \div 231 = 955.7$ gallons, nearly. Ans.

EXAMPLES FOR PRACTICE.

1. Find the convex surface of the frustum of a square pyramid, one edge of whose lower base is 15 inches long, one edge of whose upper base is 14 inches long, and whose slant height is one inch.
Ans. 58 sq. in.

2. Find the volume of the above frustum, supposing its altitude to be 3 inches. Ans. 631 cu. in.

3. Find the volume of the frustum of a cone whose altitude is 12 feet and the diameters of whose upper and lower bases are 8 and 10 feet, respectively. Ans. 766.55 cu. ft.

4. If a tank had the dimensions of example 3, how many gallons would it hold? Ans. 5,734.2 gallons, nearly.

THE SPHERE AND CYLINDRICAL RING.

428. A **sphere** (Fig. 59) is a solid bounded by a uniformly curved surface, every point of which is equally distant from a point within, called the center.

The word **ball,** or **globe,** is generally used instead of sphere.

FIG. 59.

429. To find the area of the surface of a sphere:

Rule 68.—*Square the diameter and multiply the result by 3.1416.*

EXAMPLE.—What is the area of the surface of a sphere whose diameter is 14 inches?

SOLUTION.—Diameter squared $\times 3.1416 = 14^2 \times 3.1416 = 14 \times 14 \times 3.1416 = 615.75$ sq. in. Ans.

From this it will be seen that the surface of a sphere equals the circumference of a great circle multiplied by the diameter, a rule often used; a *great circle* of a sphere is the intersection of its surface with a plane passing through its center; for instance, the *great circle* of a sphere 6 in. diameter is a circle of 6 in. diameter. Any number of *great circles* could be described on a given sphere.

430. To find the volume of a sphere:

Rule 69.—*Cube the diameter and multiply the result by .5236.*

EXAMPLE.—What is the weight of a lead ball 12 inches in diameter, a cubic inch of lead weighing .41 pound?

SOLUTION.— Diameter cubed $\times$ .5236 $= 12 \times 12 \times 12 \times$ .5236 $= 904.78$ cu. in., or the volume of the ball. The weight, therefore, $= 904.78 \times .41 = 370.96$ pounds. Ans.

431. To find the convex area of a cylindrical ring:

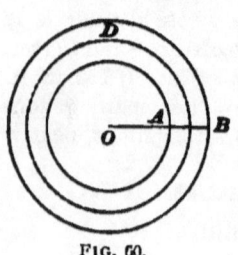

FIG. 60.

A **cylindrical ring** (Fig. 60) is a cylinder bent to a circle. The **altitude** of the cylinder before bending is the same as the length of the dotted center line D. The **base** will correspond to a cross-section on the line $A\,B$ drawn from the center O. Hence, to find the convex area:

Rule 70.—*Multiply the circumference of an imaginary cross-section on the line $A\,B$, by the length of the center line D.*

EXAMPLE.—If the outside diameter of the ring is 12 inches, and the inside diameter is 8 inches, what is its convex area?

SOLUTION.—The diameter of the center circle equals one-half the sum of the inside and outside diameters $= \dfrac{12+8}{2} = 10$, and $10 \times 3.1416 = 31.416$ inches, the length of the center line.

The radius of the inner circle is 4 inches; of the outside circle, 6 inches; therefore, the diameter of the cross-section on the line $A\,B$ is 2 inches. Then, $2 \times 3.1416 = 6.2832$ inches, and $6.2832 \times 31.416 = 197.4$ sq. in., the convex area. Ans.

MENSURATION.

432. To find the volume of a cylindrical ring:

Rule 71.—*The volume will be the same as that of a cylinder whose altitude equals the length of the dotted center line D, and whose base is the same as a cross-section of the ring on the line A B, drawn from the center O. Hence, to find the volume of a cylindrical ring, multiply the area of an imaginary cross-section on the line A B, by the length of the center line D.*

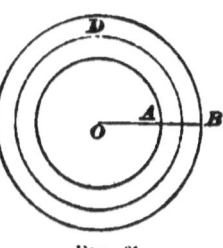

FIG. 61.

EXAMPLE.—What is the volume of a cylindrical ring whose outside diameter is 12 inches, and whose inside diameter is 8 inches?

SOLUTION.—The diameter of the center circle equals one-half the sum of the inside and outside diameters $= \dfrac{12+8}{2} = 10$.

$10 \times 3.1416 = 31.416$ inches, the length of the center line.

The radius of the outside circle $= 6$ inches; of the inside circle $= 4$ inches; therefore, the diameter of the cross-section on the line $A B = 2$ inches.

Then, $2^2 \times .7854 = 3.1416$ sq. in., the area of the imaginary cross-section.

And $3.1416 \times 31.416 = 98.7$ cubic inches, the volume. Ans.

EXAMPLES FOR PRACTICE.

1. What is the volume of a sphere 30 inches in diameter?
   Ans. 14,137.2 cu. in.

2. How many square inches in the surface of the above sphere?
   Ans. 2,827.44 sq. in.

3. Required the area of the convex surface of a circular ring, the outside diameter of the ring being 10 inches, and the inside diameter 7¼ inches. Ans. 107.95 sq. in.

4. Find the cubical contents of the ring in the last example.
   Ans. 33.73 cu. in.

5. The surface of a sphere contains 314.16 square inches. What is the volume of the sphere? Ans. 523.6 cu. in.

MECHANICS.

433. Mechanics is that science which treats of the action of forces upon bodies, and the effects which they produce; it treats of the laws which govern the movement and equilibrium of bodies, and shows how they may be utilized.

MATTER AND ITS PROPERTIES.

434. Matter is anything that occupies space. It is the substance of which all bodies consist. Matter is composed of *molecules* and *atoms*.

435. A molecule is the smallest portion of matter that can exist without changing its nature.

436. An **atom** is an indivisible portion of matter.

Atoms unite to form molecules, and a collection of molecules forms a mass or body.

A drop of water may be divided and subdivided, until each particle is so small that it can only be seen by the most powerful microscope, but each particle will still be water.

Now, imagine the division to be carried on still further, until a limit is reached beyond which it is impossible to go without changing the nature of the particle. The particle of water is now so small that, if it be divided again, it will cease to be water, and will be something else; we will call this particle a *molecule*.

If a molecule of water be divided, it will yield two atoms of hydrogen gas, and one of oxygen gas. If a molecule of sulphuric acid be divided, it will yield two atoms of hydrogen, one of sulphur, and four of oxygen.

It has been calculated that the diameter of a molecule is larger than $\frac{1}{125000000}$ of an inch, and smaller than $\frac{1}{500000000}$ of an inch.

437. Bodies are composed of collections of molecules. Matter exists in three conditions or forms: *solid*, *liquid*, and *gaseous*.

438. A solid body is one whose molecules change their relative positions with great difficulty; as iron, wood, stone, etc.

439. A liquid body is one whose molecules tend to change their relative positions easily. Liquids readily adapt themselves to the shape of vessels which contain them, and their upper surface always tends to become perfectly level. Water, mercury, molasses, etc., are liquids.

440. A gaseous body, or gas, is one whose molecules tend to separate from one another; as air, oxygen, hydrogen, etc.

Gaseous bodies are sometimes called **aeriform** (air-like) **bodies**. They are divided into two classes: the so-called "*permanent*" *gases*, and *vapors*.

441. A permanent gas is one which remains a gas at ordinary temperatures and pressures.

442. A vapor is a body which at ordinary temperatures is a liquid or solid, but when heat is applied, becomes a gas, as steam.

One body may be in all three states; as, for example, mercury, which at ordinary temperatures is a liquid, becomes a solid (freezes) at 40° below zero, and a vapor (gas) at 600° above zero. By means of great cold, all gases, even hydrogen, have been liquefied, and some solidified.

By means of heat, all solids have been liquefied, and a great many vaporized. It is probable that, if we had the means of producing sufficiently great extremes of heat and cold, all solids might be converted into gases, and all gases into solids.

MECHANICS.

443. Every portion of matter possesses certain qualities called *properties*. Properties of matter are divided into two classes: *general* and *special*.

444. General properties of matter are those which are common to all bodies. They are as follows: *Extension, impenetrability, weight, indestructibility, inertia, mobility, divisibility, porosity, compressibility, expansibility,* and *elasticity.*

445. Extension is the property of occupying space. Since all bodies must occupy space, it follows that extension is a general property.

446. By **impenetrability** we mean that no two bodies can occupy exactly the same space at the same time.

447. Weight is the measure of the earth's attraction upon a body. All bodies have weight. In former times it was supposed that gases had no weight, since, if unconfined, they tend to move away from the earth, but, nevertheless, they will finally reach a point beyond which they can not go, being held in suspension by the earth's attraction. Weight is measured by comparison with a standard. The standard is a bar of platinum owned and kept by the Government; it weighs one pound.

448. Inertia means that a body can not put itself in motion nor bring itself to rest. To do either it must be acted upon by some force.

449. Mobility means that a body can be changed in position by some force acting upon it.

450. Divisibility is that property of matter on account of which a body may be separated into parts.

451. Porosity is the term used to denote the fact that there is space between the molecules of a body. The molecules of a body are supposed to be spherical, and, hence, there is space between them, as there would be between peaches in a basket. The molecules of water are larger than those of salt; so that when salt is dissolved in water, its molecules wedge themselves between the molecules of the water, and unless too much salt is added, the water will occupy no more space than it did before. This does not prove that water is penetrable, for the molecules of salt occupy the space that the molecules of water did not.

Water has been forced through iron by pressure, thus proving that iron is porous.

452. Compressibility.—This property is a natural consequence of the preceding one. Since there is space between the molecules, it is evident that by means of force (pressure) they can be brought closer together, and thus the body be made to occupy a smaller space.

453. Expansibility is the term used to denote the fact that the molecules of a body will, under certain conditions (when heated, for example), move farther apart, and so cause the body to *expand*, or occupy a greater space.

454. Elasticity is that property of matter which enables a body when distorted within certain limits to resume its original form when the distorting force is removed. Glass, ivory, and steel are very elastic, clay and putty in their natural state being very slightly so.

455. Indestructibility is the term used to denote the fact that we can not *destroy* matter. A body may undergo thousands of changes, be resolved into its molecules, and its molecules into atoms, which may unite with other atoms to form other molecules and bodies entirely different in appearance and properties from the original body, but the same number of atoms remain. The whole number of atoms in the universe is exactly the same now as it was millions of years ago, and will always be the same. *Matter is indestructible.*

456. Special properties are those which are not possessed by all bodies. Some of the most important are as follows: *hardness, tenacity, brittleness, malleability,* and *ductility*.

457. Hardness.—A piece of copper will scratch a piece of wood, steel will scratch copper, and tempered steel will scratch steel in its ordinary state. We express all this by saying that steel is *harder* than copper, and so on. Emery and corundum are extremely hard, and the diamond is the hardest of all known substances. It can only be polished with its own powder.

458. Tenacity is the term applied to the power with which some bodies resist a force tending to pull them apart. Steel is very tenacious.

459. Brittleness.—Some bodies possess considerable power to resist either a pull or a pressure, but they are easily broken when subjected to shocks or jars; for example, good glass will bear a *greater* compressive force than most woods, but may be easily broken when dropped on to a hard floor; this property is called *brittleness*.

460. Malleability is that property which permits of some bodies being hammered or rolled into sheets. Gold is the most malleable of all substances.

461. Ductility is that property which enables some bodies to be drawn into wire. Platinum is the most ductile of all substances.

MOTION AND VELOCITY.

462. Motion is the opposite of rest, and indicates a changing of position in relation to some object which is for that purpose regarded as being fixed. If a large stone is rolled down hill, it is in motion in relation to the hill.

If a person is on a railway train, and walks in the opposite direction from that in which the train is moving, and with the same speed, he will be in motion as regards the train, but at rest with respect to the earth, since, until he gets to the end of the train, he will be directly over the spot at which he was when he started to walk.

463. The **path** of a body in motion is the line described by a certain point in the body called its center of gravity. No matter how irregular the shape of the body may be, nor how many turns and twists it may make, the line which indicates the direction of this point for every instant that it is in motion is the path of the body.

464. Velocity is rate of motion. It is measured by a unit of space passed over in a unit of time. When equal spaces are passed over in equal times, the velocity is said to be **uniform**. In all other cases it is **variable**.

If the fly-wheel of an engine keeps up a constant speed of a certain number of revolutions per minute, the velocity of any point is uniform. A railway train having a constant speed of 40 miles per hour, moves 40 miles every hour, or $\frac{40}{60} = \frac{2}{3}$ of a mile every minute, and since equal spaces are passed over in equal times, the velocity is uniform.

465. To find the uniform velocity which a body must have to pass over a certain distance or space in a given time:

Rule 72.—*Divide the distance by the time.*

EXAMPLE.—The piston of a steam engine travels 3,000 feet in 5 minutes; what is its velocity in feet per minute?

SOLUTION.—Here 3,000 feet is the distance, and 5 minutes is the time. Applying the rule, 3,000 ÷ 5 = 600 feet per minute. Ans.

CAUTION.—Before applying the above or any of the succeeding rules, care must be taken to reduce the values given to the denominations required in the answer. Thus, in the above example, had the velocity been required in feet per second instead of feet per minute, the 5 minutes would have been reduced to seconds before dividing. The operation would then have been, 5 min. = 5 × 60 = 300 sec. Applying the rule, 3,000 ÷ 300 = 10 ft. per sec. Ans.

466. Had the velocity been required in inches per second, it would have been necessary to reduce the 3,000 feet to inches and the 5 minutes to seconds, before dividing. Thus, 3,000 ft. × 12 = 36,000 in. 5 min. × 60 = 300 sec. Now, applying the rule, 36,000 ÷ 300 = 120 in. per sec. Ans.

EXAMPLE.—A railroad train travels 50 miles in 1¼ hours; what is its average velocity in feet per second?

MECHANICS. 155

SOLUTION.—Reducing the miles to feet, and the hours to seconds, 50 miles × 5,280 = 264,000 ft. 1¼ hours × 60 × 60 = 5,400 sec. Applying the rule, 264,000 ÷ 5,400 = 48⅔ ft. per sec. Ans.

467. If the uniform velocity (or the average velocity) and the time are given, and it is required to find the distance which a body having the given velocity would travel in the given time:

Rule 73.—*Multiply the velocity by the time.*

EXAMPLE.—The velocity of sound in still air is 1,092 feet per second; how many miles will it travel in 16 seconds?

SOLUTION.—Reducing the 1,092 ft. to miles, $1,092 \div 5,280 = \frac{1,092}{5,280}$. Applying the rule, $\frac{1,092}{5,280} \times 16 = 3.31$ miles, nearly. Ans.

EXAMPLE.—The piston speed of an engine is 11 ft. per sec., how many miles does the piston travel in 1 hour and 15 minutes?

SOLUTION.—1 hour and 15 minutes reduced to seconds = 4,500 seconds = the time. 11 feet reduced to miles = $\frac{11}{5,280}$ mile = velocity in miles per second. Applying the rule, $\frac{11}{5,280} \times 4,500 = 9.375$ miles. Ans.

468. If the distance through which a body moves is given, and also its average or uniform velocity, and it is desired to know how long it takes the body to move through the given distance:

Rule 74.—*Divide the distance, or space passed over, by the velocity.*

EXAMPLE.—Suppose that the radius of the crank of a steam engine is 15″, and that the shaft makes 120 revolutions per minute, how long will it take the crank-pin to travel 18,849.6 feet?

SOLUTION.—Since the radius, or distance from the center of the shaft to the center of the crank-pin, is 15″, the diameter of the circle it moves in is 15″ × 2 = 30″ = 2.5 ft. The circumference of this circle is 2.5 × 3.1416 = 7.854 ft. 7.854 × 120 = 942.48 ft. = distance that the crank-pin travels in one minute = velocity in feet per minute. Applying the rule, 18,849.6 ÷ 942.48 = 20 minutes. Ans.

EXAMPLE.—A point on the rim of an engine fly-wheel travels at the rate of 150 feet per second; how long will it take to travel 45,000 feet?

SOLUTION.—Applying the rule,
45,000 ÷ 150 = 300 seconds = 5 minutes. Ans.

EXAMPLES FOR PRACTICE.

1. A locomotive has drivers 80" in diameter. If they make 293 revolutions per minute, what is the velocity of the train in (*a*) feet per second? (*b*) miles per hour? Ans. $\begin{cases} (a)\ 102.277\ \text{ft. per sec.} \\ (b)\ \ \ 69.733\ \text{mi. per hr.} \end{cases}$

2. Assuming the velocity of steam as it enters the cylinder to be 900 feet per second, how far could it travel, if unobstructed, during the time the fly-wheel of an engine revolved 7 times, if the number of revolutions per minute were 120? Ans. 3,150 ft.

3. The average speed of the piston of an engine is 528 feet per minute, how long will it take the piston to travel 4 miles?
Ans. 40 min.

4. A speed of 40 miles per hour equals how many feet per second?
Ans. $58\frac{2}{3}$ ft.

5. The earth turns around once in 24 hours. If the diameter be taken as 8,000 miles, what is the velocity of a point on the earth in miles per minute? Ans. $17.45\frac{1}{4}$ mi. per min.

6. The stroke of an engine is 28 inches. If the engine makes 11,400 strokes per hour, (*a*) what is its speed in feet per minute? (*b*) How far will this piston travel in 11 minutes? Ans. $\begin{cases} (a)\ 443\frac{1}{3}\ \text{ft. per min.} \\ (b)\ 4{,}876\ \text{ft. 8 in.} \end{cases}$

FORCE.

469. Force is that which produces, or tends to produce or destroy, motion. Forces are called by various names, according to the effects which they produce upon a body, as *attraction, repulsion, cohesion, adhesion, accelerating* force, *retarding* force, *resisting* force, etc., but all are equivalent to a push or pull, according to the direction in which they act upon a body. That the effect of a force upon a body may be compared with another force, it is necessary that three conditions be fulfilled in regard to both bodies. They are as follows:

(1) *The point of application, or point at which the force acts upon the body, must be known.*

(2) *The direction of the force, or, what is the same thing, the straight line along which the force tends to move the point of application, must be known.*

MECHANICS.

(3) *The magnitude or value of the force, when compared with a given standard, must be known.*

470. *The unit of magnitude of forces will always be taken as one pound*, and all forces will be spoken of as a certain number of pounds.

In practice, force is always regarded as a pressure; that is, a force may always be replaced by an equivalent weight. Thus, a force of 20 lb. acting upon a body is regarded as a pressure of 20 lb. produced by a weight of 20 lb. The tendency of a force is always to produce motion in the direction in which it acts. The resistance may be too great for it to cause motion, but it *always tends* to produce it.

471. The fundamental principles of the relations between force and motion were first stated by Sir Isaac Newton. They are called "Newton's Three Laws of Motion," and are as follows:

(*1*) *All bodies continue in a state of rest, or of uniform motion in a straight line, unless acted upon by some external force that compels a change.*

(*2*) *Every motion or change of motion is proportional to the acting force, and takes place in the direction of the straight line along which the force acts.*

(*3*) *To every action there is always opposed an equal and contrary reaction.*

472. In the first law of motion it is stated that a body once set in motion by any force, no matter how small, will move forever in a straight line, and always with the same velocity, unless acted upon by some other force which compels a change. It is not possible to actually verify this law, on account of the earth's attraction for all bodies, but, from astronomical observations, we are certain that the law is true. This law is often called *the law of inertia.*

473. The word **inertia** is so abused that a full understanding of its meaning is necessary. Inertia is not a force, although it is often so called. If a force acts upon a body and puts it in motion, the effect of the force is stored in the body, and a second body, in stopping the first, will receive a blow equal in every respect to the original force, assuming that there has been no resistance of any kind to the motion of the first body.

It is dangerous for a person to jump from a fast moving train, for the reason that, since his body has the same velocity as the train, it has the same force stored in it that would cause a body of the same weight to take the same velocity as the train, and the effect of a sudden stoppage is the same as the effect of a blow necessary to give the person that velocity.

By "bracing" himself and jumping in the same direction that the train is moving, and running, he brings himself gradually to rest, and thus reduces the danger. If a body is at rest, it must be acted upon by a force in order to be put in motion, and, no matter how great the force may be, it can not be *instantly* put in motion.

The resistance thus offered to being put in motion is commonly, but erroneously, called the *"Resistance of Inertia."* It should be called the *Resistance due to Inertia*.

474. From the second law, we see that if two or more forces act upon a body, their final effect upon the body will be in proportion to their magnitudes, and to the directions in which they act.

Thus, if the wind be blowing due West, with a velocity of 50 miles per hour, and a ball be thrown due North with the same velocity, or 50 miles per hour, the wind will carry the ball West while the force of the throw is carrying it North, and the combined effect will be to cause it to move Northwest.

The amount of departure from due North will be proportional to the force of the wind, and independent of the velocity due to the force of the throw.

MECHANICS.

475. In Fig. 62 a ball e is supported in a cup, the bottom of which is attached to the lever o in such a manner that o 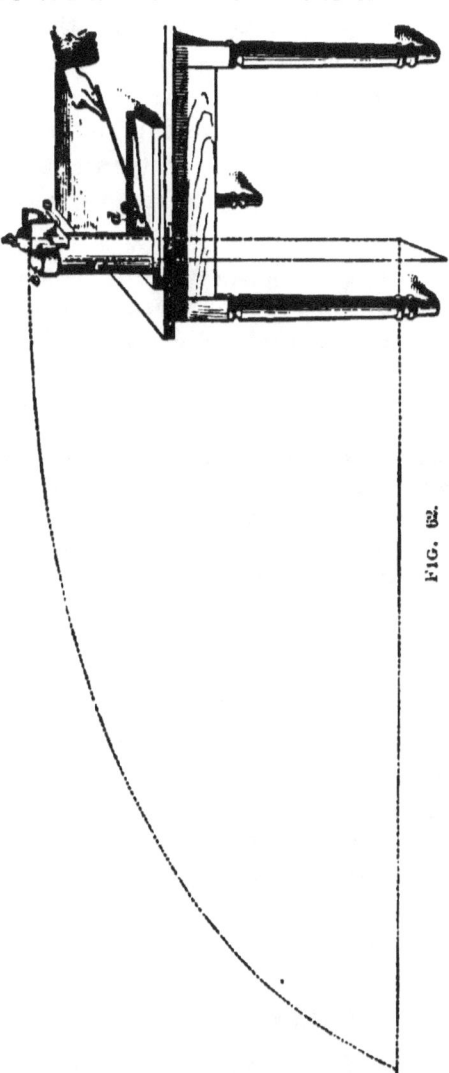 will swing the bottom horizontally and allow the ball to drop. Another ball b rests in a horizontal groove that is provided with a slit in the bottom. A swinging arm is actuated by the spring d in such a manner that, when drawn back, as shown, and then released, it will strike the lever o and the ball b at the same time. This gives b an impulse in a horizontal direction, and swings o so as to allow e to fall.

On trying the experiment, it is found that b follows a path shown by the curved dotted line, and reaches the floor at the same instant as e, which drops vertically. This shows that the force which gave the first ball its horizontal movement had no effect on the vertical force which compelled both balls to fall to the floor; the vertical force producing the same effect as if the horizontal force had not acted. The second law may also be stated as follows: *A force has the same effect in producing motion, whether it acts upon a body at rest or in motion, and whether it acts alone or with other forces.*

476. The third law states that action and reaction are equal and opposite. A man can not lift himself by his bootstraps, for the reason that he presses downwards with the same force that he pulls upwards; the downward reaction equals the upward action, and is opposite to it.

In springing from a boat we must exercise caution or the reaction will drive the boat from the shore. When we jump from the ground, we tend to push the earth from us, while the earth reacts and pushes us from it.

EXAMPLE.—Two men pull on a rope in opposite directions, each exerting a force of 100 pounds; what is the force which the rope resists?

SOLUTION.—Imagine the rope to be fastened to a tree, and one man to pull with a force of 100 pounds. The rope evidently resists 100 pounds. According to Newton's third law, the reaction of the tree is also 100 pounds. Now, suppose the rope to be slackened, but that one end is still fastened to the tree, and the second man to take hold of the rope near the tree, and pull with a force of 100 pounds, the first man pulling as before. The resistance of the rope is 100 pounds, as before, since the second man merely takes the place of the tree. *He is obliged to exert a force of 100 pounds to keep the rope from slipping through his fingers.* If the rope be passed around the tree, and each man pulls an end with a force of 100 pounds in the same and parallel directions, the stress in the rope is 100 pounds, as before, but the tree must resist the pull of both men, or 200 pounds.

477. A **force** may be represented by a line; thus, in Fig. 63, let A be the *point of application* of the force; let the length of the line AB represent its *magnitude*, and let the arrowhead indicate the *direction* in which the force acts, then the line AB fulfils the three conditions (see Art. **469**), and the force is fully represented.

FIG. 63.

CENTER OF GRAVITY.

478. *The center of gravity of a body is that point at which the body may be balanced, or it is the point at which the whole weight of a body may be considered as concentrated.*

This point is not always *in* the body; in the case of a horseshoe or a ring it lies outside of the substance of, but within the space enclosed by, the body.

MECHANICS. 161

In a moving body, the line described by its center of gravity is always taken as the path of the body. In finding the distance that a body has moved, the distance that the center of gravity has moved is taken.

The definition of the center of gravity of a body may be applied to a system of bodies if they are considered as being connected at their centers of gravity.

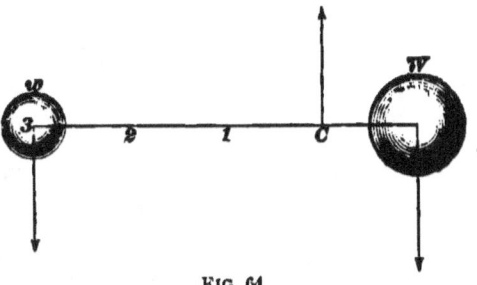

FIG. 64.

If w and W, Fig. 64, be two bodies of known weight, their center of gravity will be at C. The point C may be readily determined as follows:

Rule 75.—*The distance of the common center of gravity, from the center of gravity of the large weight, is equal to the weight of the smaller body multiplied by the distance between the centers of gravity of the two bodies, and this product divided by the sum of the weights of the two bodies.*

EXAMPLE.—In Fig. 64, $w = 10$ pounds, $W = 30$ pounds, and the distance between their centers of gravity is 36 inches where is the center of gravity of both bodies situated?

SOLUTION.—Applying the rule, $10 \times 36 = 360$. $10 + 30 = 40$. $360 \div 40 = 9" =$ distance of center of gravity from center of large weight.

479. It is now very easy to extend this principle, to find the center of gravity of any number of bodies when their weights and the distances apart of their centers of gravity are known, by the following rule:

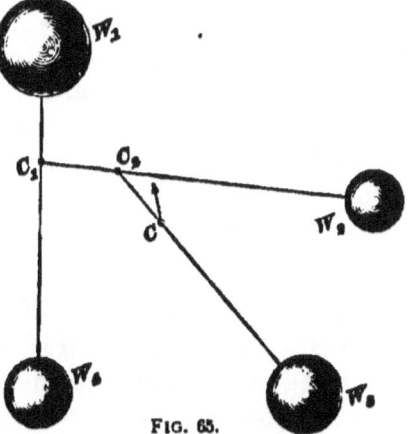

FIG. 65.

Rule 76.—*Find the center of gravity of two of the bodies as W_1 and W_4, in Fig. 65. Assume that the weight of both* bodies is concentrated at C_1, and find the center of gravity of this combined weight at C_1, and the weight of W_3; let it be at C_2; then find the center of gravity of the combined weights of W_1, W_2, W_3 (concentrated at C_2), and W_4; let it be at C; then C will be the center of gravity of the four bodies.

480. To find the center of gravity of **any parallelogram:**

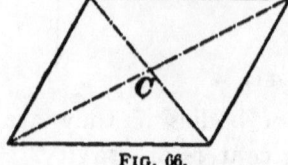

Fig. 66.

Rule 77.—*Draw the two diagonals, Fig. 66, and their point of intersection C will be the center of gravity.*

481. To find the center of gravity of a **triangle,** as $A B C$, Fig. 67:

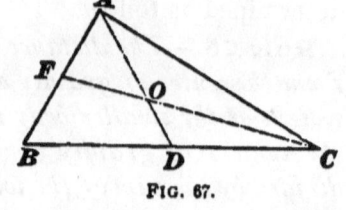

Fig. 67.

Rule 78.—*From any vertex, as A, draw a line to the middle point D of the opposite side $B C$. From one of the other vertexes, as C, draw a line to F, the middle point of the opposite side $A B$; the point of intersection O of these two lines is the center of gravity.*

It is also true that the distance $D O = \frac{1}{3} D A$, and that $F O = \frac{1}{3} F C$, and the center of gravity could have been found by drawing from any vertex a line to the middle point of the opposite side, and measuring back from that side $\frac{1}{3}$ of the length of the line.

The center of gravity of **any regular plane figure** is the same as the center of the inscribed or circumscribed circle.

482. To find the center of gravity of **any irregular plane figure,** but of uniform thickness throughout, divide one of the parallel surfaces into triangles, parallelograms, circles, ellipses, etc., according to the shape of the figure; find the area and center of gravity of each part separately, and combine the centers of gravity thus found in the same manner as in rule **76;** in this case, however, dealing with the *area* of each part instead of its weight. See Fig. 68.

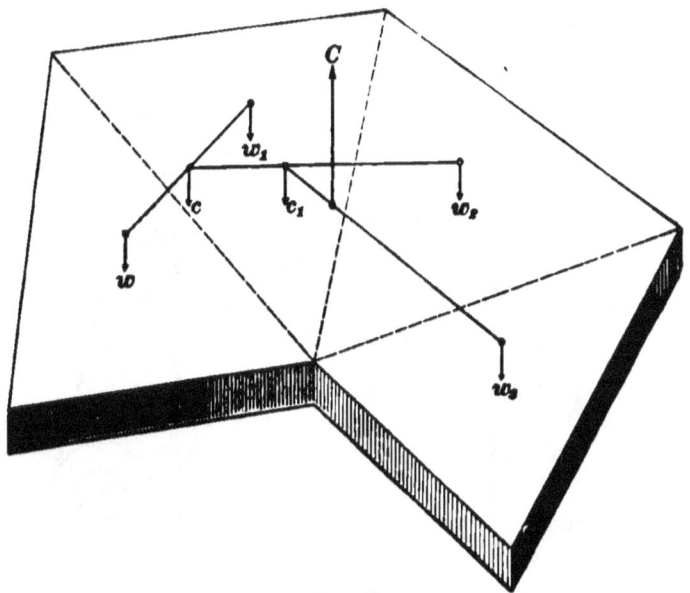

FIG. 68.

EXAMPLE.—Suppose that the two balls shown in Fig. 64 were 5" and 10" in diameter, and weighed 10 lb. and 80 lb., respectively. If the distance between their centers were 40", and they were connected by a steel rod 1" in diameter, where would the center of gravity be, taking the weight of a cubic inch of steel as .283 lb.?

SOLUTION.—The length of the rod $= 40 - \frac{5}{2} - \frac{10}{2} = 32\frac{1}{2}"$. Its volume is $1^2 \times .7854 \times 32\frac{1}{2} = 25.53$ cu. in. $25.53 \times .283 = 7.22$ lb. The rod being straight, its center of gravity is in the middle at a distance of $\frac{32.5}{2} + \frac{5}{2} = 18\frac{3}{4}"$ from the center of the smaller weight, and $\frac{32.5}{2} + \frac{10}{2} = 21\frac{1}{4}"$ from the center of the larger weight. Now, considering the weight of the rod to be concentrated at its center of gravity, we have three weights of 10, 7.22, and 80 lb., all in a straight line, and the distances between them given to find the center of gravity, or balancing point, of the combination, by rule **76**. We will first find the center of gravity of the two smaller weights by rule **75**, as follows: $7.22 \times 18\frac{3}{4} = 135.38$. $10 + 7.22 = 17.22$. $135.38 \div 17.22 = 7.86"$ = distance from the center of the 10-lb. weight. Considering both of the smaller weights to be concentrated at this point, we find the center of gravity of this combined weight and the large weight as follows. $40 - 7.86 = 32.14"$ = distance between the center of gravity of the two small weights and the center of gravity of the 80-lb. weight. Applying rule **75**, 17.22 × 32.14 = 553.45.

17.22 + 80 = 97.22. 553.45 ÷ 97.22 = 5.693′ = distance from the center of the 80-lb. weight. Ans.

483. Center of Gravity of a Solid.—In a body free to move, the center of gravity will lie in a vertical plumb

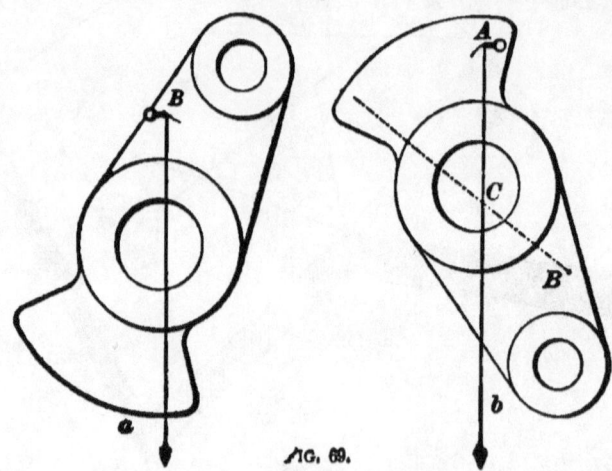

FIG. 69.

line drawn through the point of support. Therefore, to find the position of the center of gravity of an irregular solid, as the crank, Fig. 69, suspend it at some point, as B, so that it will move freely. Drop a plumb line from the point of suspension, and mark its direction. Suspend the body at another point, as A, and repeat the process. The intersection C of the two lines will be directly over the center of gravity.

Since the center of gravity depends wholly upon the shape and weight of a body, it may be without the body, as in the case of a circular ring whose center of gravity is the same as the center of the circumference of the ring.

EXAMPLES FOR PRACTICE.

1. A spherical shell has a wrought-iron handle attached to it. The shell is 10″ in diameter, and weighs 20 lb. The handle is 1¼″ in diameter, and the distance from the center of the shell to the end of the handle is 4 ft. Where is the center of gravity? Take the weight of a cubic inch of wrought iron as .278 lb.   Ans. 13.612″ from center of shell.

MECHANICS. 165

2. The distance between the centers of two bodies is 51'. The weights of the bodies being 20 and 73 lb., where is the center of gravity?
Ans. 10.968' from the center of large weight.

3. A hollow engine piston weighs 275 lb., and is $3\frac{1}{4}$' thick. Assuming the piston rod to be straight throughout its entire length, and to weigh 140 lb., at what point will the piston and rod balance if the length of the rod is 73' from the face of the piston? Consider the weight of the piston to be concentrated at its center.
Ans. 11.15', nearly, from face of piston.

4. Weights of 5, 9, and 12 lb. lie in one straight line in the order named. Distance from the 5-lb. weight to the 9-lb. weight is 22', and from the 9-lb. weight to the 12-lb. weight is 18'. Where is the center of gravity?
Ans. 13.923' from 12-lb. weight.

SIMPLE MACHINES.

484. A **lever** is a bar capable of being turned about a pivot, or point, as in Figs. 70, 71, and 72.

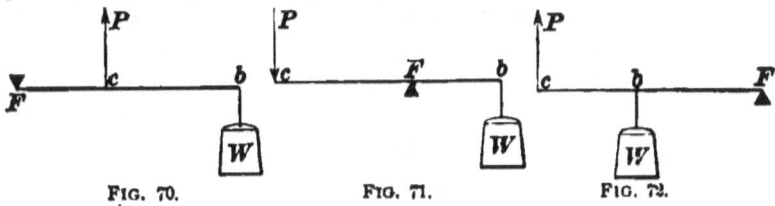

FIG. 70.   FIG. 71.   FIG. 72.

The object W to be lifted is called the **weight**; the force employed P is called the **power**; and the point, or pivot, F is called the **fulcrum**.

485. That part of the lever between the fulcrum and the weight, or $F b$, is called the **weight arm**, and the part between the fulcrum and the power, or $F c$, is called the **power arm**.

In order that the lever shall be in equilibrium (balance), *the power multiplied by the power arm must equal the weight multiplied by the weight arm;* that is, $P \times F c$ must equal $W \times F b$.

486. If F be taken as the center of a circle, and arcs be described through b and c, it will be seen that if the weight arm be moved through a certain angle, the power arm will move through the same angle. Since, in the same or equal angles, the lengths of the arcs are proportional to the radii with which they were described, it is seen that the power arm is proportional to the distance through which the power moves, and the weight arm is proportional to the distance through which the weight moves.

Hence, instead of writing $P \times Fc = W \times Fb$, we might have written it $P \times$ distance through which P moves $= W \times$ distance through which W moves. This is the general law of all machines, and can be applied to any mechanism from the simple lever up to the most complicated arrangement. Stated in the form of a rule, it is as follows:

Rule 79.—*The power multiplied by the distance through which it moves equals the weight multiplied by the distance through which it moves.*

487. In the above rule, it will be noticed that there are four requirements necessary for a complete knowledge of the lever, viz.: the power (or force), the weight, the power arm (or distance through which the power moves), and the weight arm (or distance through which the weight moves). If any three are given, the fourth may be found by letting x represent the requirement which is to be found, and multiplying the power by the power arm and the weight by the weight arm; then, dividing the product of the two known numbers by the number by which x is multiplied; the result will be the requirement which was to be found.

EXAMPLE.—If the weight arm of a lever is 6 in. long, and the power arm is 4 ft. long, how great a weight can be raised by a force of 20 lb. at the end of the power arm?

SOLUTION.—In this example, the weight is unknown; hence, representing it by x, we have, after reducing the 4 ft. to inches, $20 \times 48 = 960 =$ power multiplied by the power arm, and $x \times 6 =$ weight multiplied by the weight arm. Dividing the 960 by 6, the result is 160 lb., the weight. Ans.

488. If the distance through which the power or weight moved had been given instead of the power arm or weight arm, and it were required to find the power or weight, the process would have been exactly the same, using the given distance instead of the power arm or weight arm.

EXAMPLE.—If, in the above example, the weight had moved $2\tfrac{1}{4}''$, how far would the power have moved?

SOLUTION.—In this example, the distance through which the power moves is required. Let x represent the distance. Then, $20 \times x =$ distance multiplied by power, and $2\tfrac{1}{2} \times 160 = 400 =$ distance multiplied by the weight. Hence, $x = \dfrac{400}{20} = 20$ in. $=$ distance through which the power arm moves.

The ratio between the weights and the power is $160 \div 20 = 8$. The ratio between the distance through which the weight moves and the distance through which the power moves is $2\tfrac{1}{2} \div 20 = \tfrac{1}{8}$. This shows that while a force of 1 lb. can raise a weight of 8 lb., the 1-lb. weight must move through 8 times the distance that the 8-lb. weight does. It will also be noticed that the ratio of the lengths of the two arms of the lever is also 8, since $48 \div 6 = 8$.

489. The law which governs the straight lever also governs the bent lever, but care must be taken to determine the true lengths of the lever arms, which are in every case *the perpendicular distances from the fulcrum to the line of direction of the weight or power.*

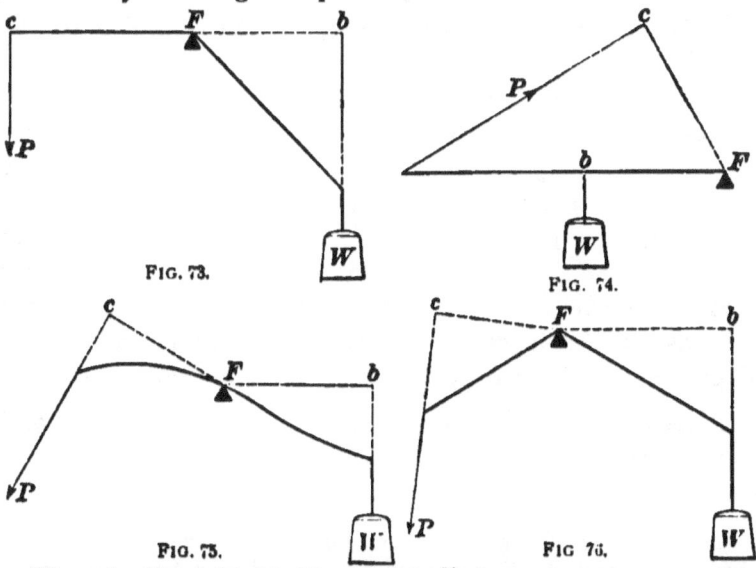

FIG. 73.   FIG. 74.
FIG. 75.   FIG. 76.

Thus, in Figs. 73, 74, 75, and 76, Fc in each case represents the power arm, and Fb the weight arm.

490. A compound lever is a series of single levers arranged in such a manner that when a power is applied to the first it is communicated to the second, and from that to the third, and so on.

Fig. 77 shows a compound lever. It will be seen that when a power is applied to the first lever at P it will be

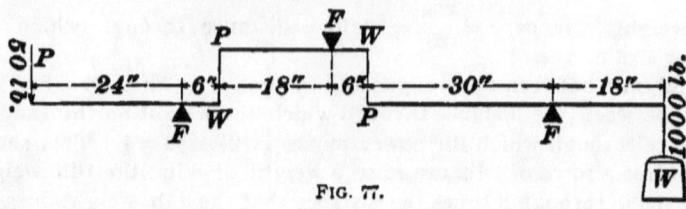

FIG. 77.

communicated to the second lever at P, from this to the third lever at P, and thus raise the weight W.

The weight which the power of the first lever could raise acts as the power of the second, and the weight which this could raise through the second lever acts as the power of the third lever, and so on, no matter how many single levers make up the compound lever.

In this case, as in every other, the power multiplied by the distance through which it moves equals the weight multiplied by the distance through which it moves.

Hence, if we move the P end of the lever, say 4 inches, and the end carrying the weight W moves $\frac{1}{5}$ of an inch, we know that the ratio between P and W is the same as the ratio between $\frac{1}{5}$ and 4; that is, 1 to 20, and, hence, that 10 pounds at P would balance 200 pounds at W, without measuring the lengths of the different lever arms. If the lengths of the lever arms are known, the ratio between P and W may be readily obtained from the following rule:

Rule 80.—*The continued product of the power and each power arm equals the continued product of the weight and each weight arm.*

EXAMPLE.—If, in Fig. 77, $PF =$ 24 inches, 18 inches, and 30 inches, respectively, and $WF =$ 6 inches, 6 inches, and 18 inches, respectively, how great a force at P would it require to raise 1,000 pounds at W? What is the ratio between W and P?

SOLUTION.—Let x represent the power; then, $x \times 24 \times 18 \times 30 =$ 12,960 $x =$ continued product of the power and each power arm. 1,000 ×

$6 \times 6 \times 18 = 648,000 =$ continued product of the weight and each weight arm, and, since $12,960\, x = 648,000$,

$$x = \frac{648,000}{12,960} = 50 \text{ lb.} = \text{the power. Ans.}$$

$1,000 \div 50 = 20 =$ ratio between W and P.

491. The **wheel and axle** consists of *two cylinders of different diameters, rigidly connected*, so that they turn together about a common axis, as in Fig. 78. Then, as

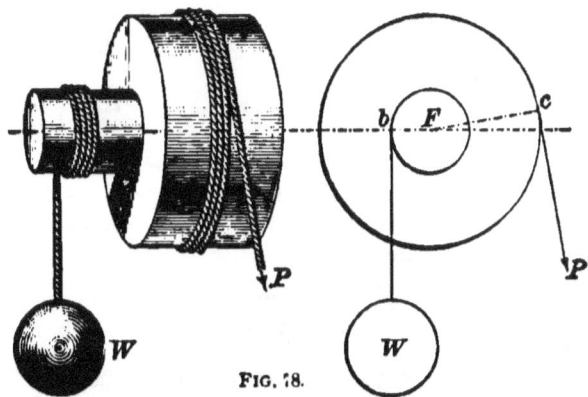

FIG. 78.

before, $P \times$ distance through which it moves $= W \times$ distance through which it moves; and, since these distances are proportional to the radii of the power cylinder and weight cylinder, $P \times Fc = W \times Fb$.

It is not necessary that an entire wheel be used; an arm, projection, radius, or anything which the power causes to revolve in a circle, may be considered as the wheel. Consequently, if it is desired to hoist a weight with a windlass, Fig. 79, the power is applied to the handle of the crank, and the distance between the center line of the crank handle and the axis of the drum corresponds to the radius of the wheel.

FIG. 79.

EXAMPLE.—If the distance between the center line of the handle and the axis of the drum, in Fig. 79, is 18 inches and the diameter of the drum is 6 inches, what force will be required at P to raise a load of 300 pounds?

SOLUTION.—$P \times (18 \times 2) = 300 \times 6$, or $P = 50$. Ans.

EXAMPLES FOR PRACTICE.

1. The lever of a safety valve is of the form shown in Fig. 70, where the force is applied at a point between the fulcrum and the weight lifted. If the distance from the fulcrum to the valve is $5\frac{1}{4}''$, and from the fulcrum to the weight is $42''$, what total force is necessary to raise the valve, the weight being 78 lb., and the weight of valve and lever being neglected? Ans. 595.64 lb.

2. If, in Fig. 77, $PF = 10''$, $12''$, $14''$, and $16''$ respectively, and $WF = 2''$, $3''$, $4''$, and $5''$, respectively, (*a*) how great a weight can a force of 20 lb. raise? (*b*) What is the ratio between W and P? (*c*) If P moves $4''$, how far will W move?

Ans. $\begin{cases} (a)\ 4{,}480\ \text{lb.} \\ (b)\ 224. \\ (c)\ \frac{1}{56}''. \end{cases}$

3. A windlass is used to hoist a weight. If the diameter of the drum on which the rope winds is $4''$, and the distance from the center of the handle to the axis of the drum is $14''$, how great a weight can a force of 32 lb. applied to the handle raise? Ans. 224 lb.

PULLEYS.

492. Pulleys for the transmission of power by belts may be divided into two principal classes: 1. The solid pulley shown in Fig. 80, in which the hub, arms, and rim are

FIG. 80.   FIG. 81.

one entire casting. 2. The split pulley shown in Fig. 81, which is cast in halves.

MECHANICS.

The latter style of pulley is more readily placed upon and removed from the shaft than the solid pulley. Pulleys are generally cast in halves or parts when they are more than six feet in diameter; this is done on account of shrinkage strain in large pulley castings, which renders them liable to crack as a result of unequal cooling of the metal.

493. Crowning.—In Fig. 82 is shown a section of the rim of a pulley that has crowning, or, in other words, whose diameter is larger at the center of the face than at its edges. This is done to prevent the belt from running off the pulley. The amount of crowning given to pulleys varies from $\frac{3}{16}$ to $\frac{1}{2}$ an inch per foot of width of the pulley face.

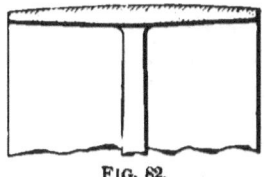

FIG. 82.

494. Balanced Pulleys.—All pulleys which rotate at high speeds should be balanced. If they are not, the centrifugal force which is generated by the pulley's rotation is greater on one side than on the other, and it will cause the pulley shaft to vibrate and shake. Pulleys should run true, so that the strain, or tension, of the belt is equal at all parts of the revolution, thus making the transmitting power equal. The smoother the surface of a pulley, the greater is its driving power.

The transmitting power of a pulley can be increased by covering the face of the pulley with a leather or rubber band; this increases the driving power about one-quarter.

495. The pulley that imparts motion to the belt is called the *driver;* that which receives the motion is called the *driven.*

The revolutions of any two pulleys over which a belt is run vary in an inverse proportion to their diameters; consequently, if a pulley of 20 inches in diameter is driven by one of 10 inches in diameter, the 20-inch pulley will make one revolution while the 10-inch pulley makes two revolutions, or they are in the ratio of 2 to 1. From this the following formulas have been deduced:

MECHANICS.

Let D = diameter of the driver;
d = diameter of the driven;
N = number of revolutions of the driver;
n = number of revolutions of the driven.

Note.—The words revolutions per minute are frequently abbreviated to R. P. M.

496. To find the diameter of the driving pulley when the diameter of the driven pulley and the number of revolutions per minute of each is given:

Rule 81.—*The diameter of the driving pulley equals the product of the diameter and number of revolutions of the driven pulley divided by the number of revolutions of the driving pulley.*

That is, $$D = \frac{dn}{N}.$$

Example.—The driving pulley makes 100 revolutions per minute, the driven pulley makes 75 revolutions per minute, and is 18 inches in diameter; what is the diameter of the driving pulley?

Solution.—Formula, $$D = \frac{dn}{N}.$$

Substituting, we have $D = \dfrac{18 \times 75}{100} = 13\frac{1}{2}$ in. Ans.

497. The diameter and number of revolutions per minute of the driving pulley being given, to find the diameter of the driven pulley, which must make a given number of revolutions per minute:

Rule 82.—*The diameter of the driven pulley equals the product of the diameter and number of revolutions of the driving pulley divided by the number of revolutions of the driven pulley.*

That is, $$d = \frac{DN}{n}.$$

Example.—The diameter of the driver is $13\frac{1}{2}$ inches, and makes 100 revolutions per minute; what must be the diameter of the driven to make 75 revolutions per minute?

Solution.—Formula, $$d = \frac{DN}{n}.$$

Substituting, we have $d = \dfrac{13\frac{1}{2} \times 100}{75} = 18$ in. Ans.

MECHANICS. 173

498. To find the number of revolutions per minute of the driven pulley, its diameter and the diameter and number of revolutions per minute of the driving pulley being given:

Rule 83.—*The number of revolutions of the driven pulley equals the product of the diameter and the number of revolutions of the driver divided by the diameter of the driven pulley.*

That is, $n = \dfrac{DN}{d}$.

EXAMPLE.—The driver is $13\frac{1}{2}$ inches in diameter, and makes 100 revolutions per minute; how many revolutions will the driven make in one minute, if it is 18 inches in diameter?

SOLUTION.—Formula, $n = \dfrac{DN}{d}$.

Substituting, we have $n = \dfrac{13\frac{1}{2} \times 100}{18} = 75$ R. P. M. Ans.

499. To find the number of revolutions per minute of the driving pulley, its diameter and the diameter and number of revolutions per minute of the driven pulley being given:

Rule 84.—*The number of revolutions of the driving pulley equals the product of the diameter and number of revolutions of the driven pulley divided by the diameter of the driving pulley.*

That is, $N = \dfrac{dn}{D}$.

EXAMPLE.—The driven pulley is 18 inches in diameter, and makes 75 revolutions per minute; how many revolutions will the driver make in one minute, if it is $13\frac{1}{2}$ inches in diameter?

SOLUTION.—Formula, $N = \dfrac{dn}{D}$.

Substituting, we have $N = \dfrac{18 \times 75}{13\frac{1}{2}} = 100$ R. P. M. Ans.

500. Wheelwork.—When there is a combination of wheels and axles, as in Fig. 83, it is called a **train.** The wheel to which the power is applied, that is, the one which imparts the motion, is called the **driver;** that which receives it, the **driven,** or **follower,** and the small wheel upon the axle, the **pinion.**

501. It will be seen that the wheel and axle bears the same relation to the train that the simple lever does to

174 MECHANICS.

the compound lever. Letting D_1, D_2, D_3, etc., represent

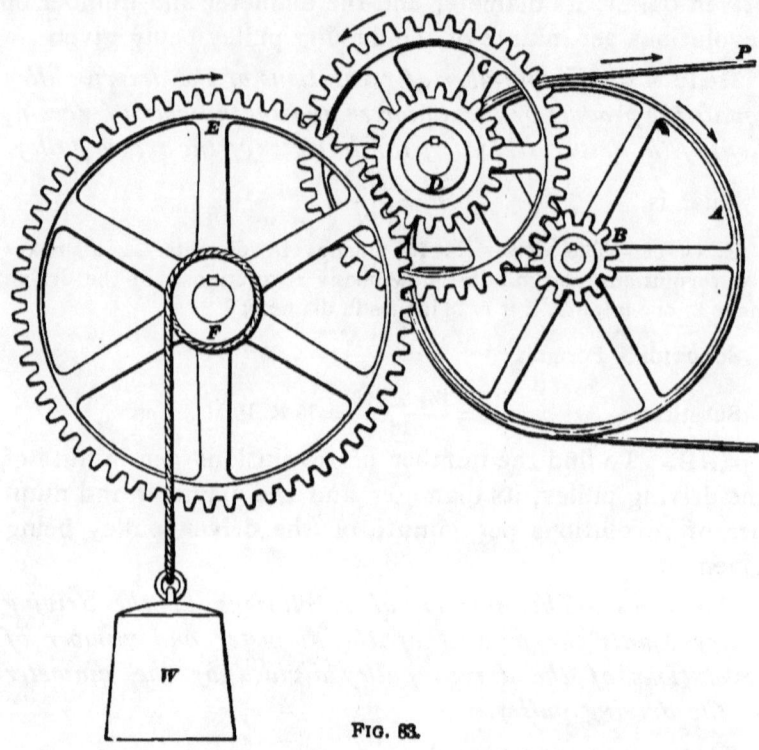

FIG. 63.

the diameters of the different drivers, and d_1, d_2, d_3, etc., the diameters of the different pinions, we have the following:

Rule 85.—*The continued product of the power and the radii of the drivers equals the continued product of the weight, the radius of the drum that moves the weight, and the radii of the pinions.*

That is, $P \times D_1 \times D_2 \times D_3$, etc., $= W \times d_1 \times d_2 \times d_3$, etc.

EXAMPLE.—If the radius of the pulley A is 20 inches; of C, 15 inches, and of E, 24 inches, and the radius of the drum F, is 4 inches; of the pinion D, 5 inches, and of the pinion B, 4 inches, how great a weight will a force of 1 pound at P raise?

SOLUTION.—Formula, $P \times D_1 \times D_2 \times D_3 = W \times d_1 \times d_2 \times d_3$.

Substituting, we have $1 \times 20 \times 15 \times 24 = W \times 4 \times 5 \times 4$, or

$$W = \frac{7{,}200}{80} = 90 \text{ lb. Ans.}$$

MECHANICS.

Hence, also, if W were raised 1 inch, P would fall 90 inches, or P would have to move through 90 inches to raise W through 1 inch.

502. It is now clear that another great law has made itself manifest, and that is that *whenever there is a gain in power without a corresponding increase in the initial force, there is a loss in speed;* this is true of any machine.

In the last example if P were to move the entire 90 inches in one second, W would move only 1 inch in one second.

Instead of using the diameter or radius of a gear, the number of teeth may be used when computing the weight which can be raised, or the velocity, as in the last example.

EXAMPLE.—The radius of the pulley A, Fig. 83, is 40', and that of F is 12". The number of teeth in B is 9; in C, 27; in D, 12, and in E, 36. If the weight to be lifted is 1,800 lb., how great a force at P is it necessary to apply to the belt?

SOLUTION.—Let P represent the force (power); then, by the rule, 85, $P \times 40 \times 27 \times 36 = 1,800 \times 12 \times 9 \times 12$, or $P \times 38,880 = 2,332,800$. Hence, $P = \dfrac{2,332,800}{38,880} = 60$ lb. = force necessary to apply to the belt.

Ans.

GEAR-WHEELS.

503. A wheel that is provided with teeth to mesh with similar teeth upon another wheel is called a **gear-wheel**, or **gear**. In Fig. 84 is shown a **spur gear**. On spur gears the teeth are always parallel to the axis of the wheel or to its shaft.

FIG. 84.

504. In Fig. 85 is shown a pair of **bevel gears** in mesh, of which one is smaller than the other.

505. When both are of the same diameter they are called **miter gears**.

In Fig. 86 is shown a pair of miter gears in mesh. It is

176   MECHANICS.

obvious that the angle which the teeth of these gears make with the axis of the shaft must be 45°.

506. In Fig. 87 is shown a revolving screw, or worm, as it is called, in gear; it is used to transmit motion from one shaft to another at right angles to it.

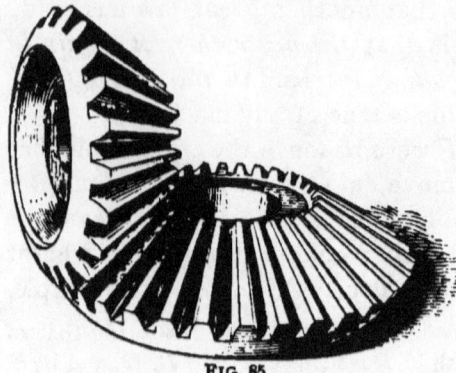

FIG. 85.

As the worm is nothing else than a screw, each revolution given to the worm will rotate the worm-wheel a distance equal to its pitch; consequently, if there are 40 teeth in the worm-wheel, a single-threaded worm will

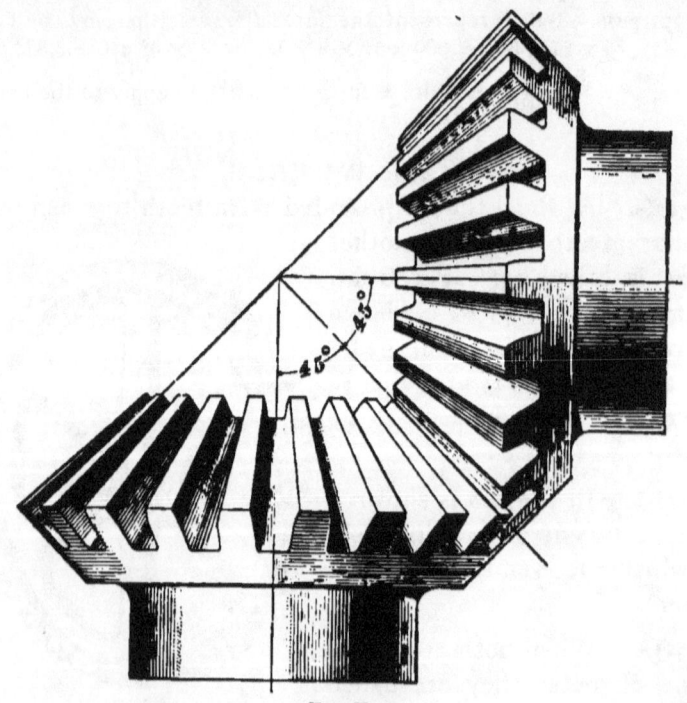

FIG. 86.

have to make 40 revolutions in order to turn the wheel once.

MECHANICS.

507. In Fig. 88 is shown a section of a rack and pinion, both having epicycloidal teeth. The arc $C\ C$ represents part of the **pitch circle**; it is on the pitch circle that all the teeth are laid out. The diameter of a gear or worm-wheel is always taken as the diameter of this circle, unless otherwise specially stated as "diameter over all," or "diameter at the root," etc.

The **pitch** of the teeth of the gear-wheel is the distance from the edge of one tooth to the corresponding edge of the following tooth measured on the pitch circle; it is marked *pitch* in the figure.

The length of the tooth of a gear-wheel is .7 of its pitch, .4 of it, called the **root**, being below or within the pitch circle, and .3 of it, called the **addendum**, being above or without the pitch circle. Thus, if the pitch of the teeth of a gear-wheel is 2 inches, the length of a tooth below the pitch circle is $2 \times .4 = .8$ of an inch; and its length above the

FIG. 87.

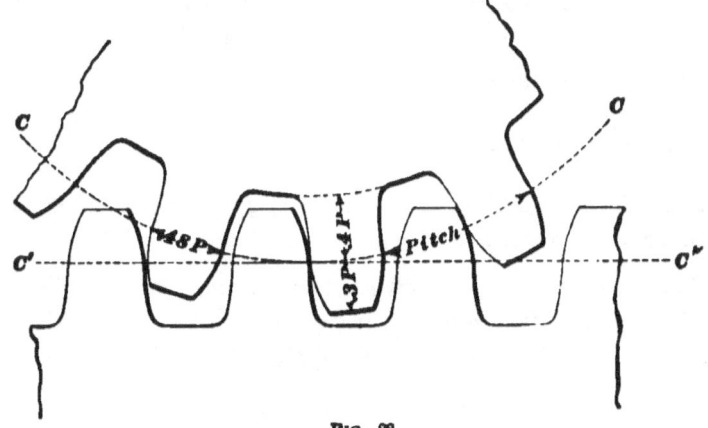

FIG. 88.

pitch circle is $2 \times .3 = .6$ of an inch. Consequently, we have only to multiply the pitch by .4 to obtain the length of the teeth below the pitch circle, and by .3 to obtain the length of the teeth above the pitch circle. The thickness of the teeth of a cast gear-wheel equals $.48 \times P$, that is, .48 of the pitch; therefore, the thickness of the above teeth is $.48 \times 2$, or .96 of an inch.

A rack may be considered as a gear-wheel rolled out so as to make the pitch circle a straight line, as $C'C''$. The teeth of racks are proportioned by the same rules as those of gear-wheels.

508. For the purpose of calculating the pitch diameter, number of teeth, etc., of the gear-wheels, we have the following rules:

Let $P =$ pitch;

$T =$ number of teeth;

$D =$ pitch diameter of the wheel.

To find the pitch diameter of a gear-wheel in inches, when the pitch and number of teeth are given:

Rule 86.—*The pitch diameter equals the product of the pitch and number of teeth divided by 3.1416.*

That is, $$D = \frac{PT}{3.1416}.$$

EXAMPLE.—What is the diameter of the pitch circle of a gear-wheel which has 75 teeth, and whose pitch is 1.675 inches?

SOLUTION.—Formula, $D = \dfrac{PT}{3.1416}.$

Substituting, we have $D = \dfrac{1.675 \times 75}{3.1416} = 40$ in. Ans.

To find the number of teeth in a gear-wheel when the diameter and pitch are given:

Rule 87.—*The number of teeth equals the product of 3.1416 and the diameter divided by the pitch.*

That is, $$T = \frac{3.1416\, D}{P}.$$

EXAMPLE.—The diameter of a gear-wheel is 40 inches, and the pitch of the teeth is 1.675 inches; how many teeth are there in the wheel?

MECHANICS. 179

SOLUTION.—Formula, $T = \dfrac{3.1416\, D}{P}$.

Substituting, we have $T = \dfrac{3.1416 \times 40}{1.675} = 75$ teeth. Ans.

To find the pitch of a gear-wheel when the diameter and the number of teeth are given.

Rule 88.—*The pitch of the teeth equals the product of 3.1416 and the diameter divided by the number of teeth.*

That is, $P = \dfrac{3.1416\, D}{T}$.

EXAMPLE.—The diameter of a gear-wheel is 40 inches, and it has 75 teeth; what is the pitch of the teeth?

SOLUTION.—Formula, $P = \dfrac{3.1416\, D}{T}$.

Substituting, we have $P = \dfrac{3.1416 \times 40}{75} = 1.675$ in. pitch. Ans.

509. The forms of teeth used in ordinary practice are the epicycloidal and involute.

Fig. 88 shows the epicycloidal form, which is composed of two different curves, the curve from the pitch circle to the top of the tooth being an epicycloid, and that from the pitch circle to the bottom of the tooth being a hypocycloid.

In gear-wheels where this form of tooth is employed, their pitch circles must run tangent to one another.

510. In Fig. 89 is shown the involute form of teeth, or

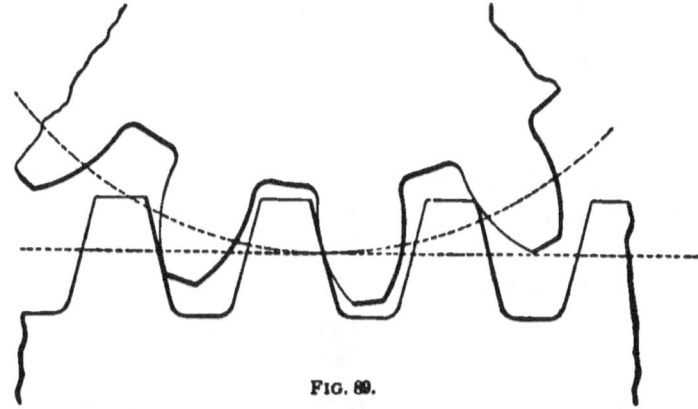

FIG. 89.

teeth having but one curve. The outlines of the teeth shown in the rack are formed of straight lines.

Involute teeth have two great advantages over epicycloidal teeth: 1. They are stronger for the same pitch, as they are thicker at the root. 2. They may be spread apart so that their pitch circles do not run tangent to one another without practically affecting the perfect action of the teeth.

511. To calculate the number of teeth or speed of one of two gear-wheels which are to gear together:

Let N = number of revolutions per minute of the driver;
n = number of revolutions per minute of the driven;
T = number of teeth in the driver;
t = number of teeth in the driven.

Rule 89.—*The number of teeth in the driver equals the product of the number of teeth and number of revolutions of the driven divided by the number of revolutions of the driver.*

That is, $$T = \frac{tn}{N}.$$

EXAMPLE.—The driven has 27 teeth, and will make 66 revolutions per minute; if the driver makes 99 revolutions per minute, how many teeth are there in the driver?

SOLUTION.—Formula, $T = \dfrac{tn}{N}.$

Substituting, we have $T = \dfrac{27 \times 66}{99} = 18$ teeth.   Ans.

The number of revolutions per minute of the driver and driven and the number of teeth in the driver being given, to find the number of teeth in the driven:

Rule 90.—*The number of teeth in the driven equals the product of the number of teeth and revolutions per minute of the driver divided by the number of revolutions per minute of the driven.*

That is, $$t = \frac{TN}{n}.$$

EXAMPLE.—The driver has 18 teeth, and makes 99 revolutions per minute, and the driven must make 66 revolutions per minute; how many teeth must there be in the driven?

SOLUTION.—Formula, $t = \dfrac{TN}{n}$.

Substituting, we have $t = \dfrac{18 \times 99}{66} = 27$ teeth. Ans.

The number of teeth in the driver and driven and the number of revolutions per minute of the driver being given, to find the number of revolutions per minute of the driven:

Rule 91.—*The number of revolutions per minute of the driven equals the product of the number of teeth and number of revolutions of the driver divided by the number of teeth of the driven.*

That is, $\qquad n = \dfrac{TN}{t}$.

EXAMPLE.—There are 18 teeth in the driver, and it makes 99 revolutions per minute; how many revolutions per minute will the driven make if it has 27 teeth?

SOLUTION.—Formula, $\qquad n = \dfrac{TN}{t}$.

Substituting, we have $n = \dfrac{18 \times 99}{27} = 66$ R. P. M. Ans.

The number of teeth in the driver and driven and the number of revolutions per minute of the driven being given, to find the number of revolutions per minute of the driver:

Rule 92.—*The number of revolutions of the driver equals the product of the number of teeth and revolutions of the driven divided by the number of teeth of the driver.*

That is, $\qquad N = \dfrac{tn}{T}$.

EXAMPLE.—If there are 27 teeth in the driven, and if it makes 66 revolutions per minute, how many revolutions per minute will the driver make if it has 18 teeth?

SOLUTION.—Formula, $\qquad N = \dfrac{tn}{T}$.

Substituting, we have $N = \dfrac{27 \times 66}{18} = 99$ R. P. M. Ans.

EXAMPLE.—In Fig. 90, the crank-shaft makes 60 revolutions per minute; the governor pulley is 4′ in diameter, the bevel gear on the governor pulley shaft has 19 teeth; the bevel gear which meshes with it and drives the governor has 30 teeth. The governor is to make 95 revolutions per minute; what should be the size of the pulley on the crank-shaft?

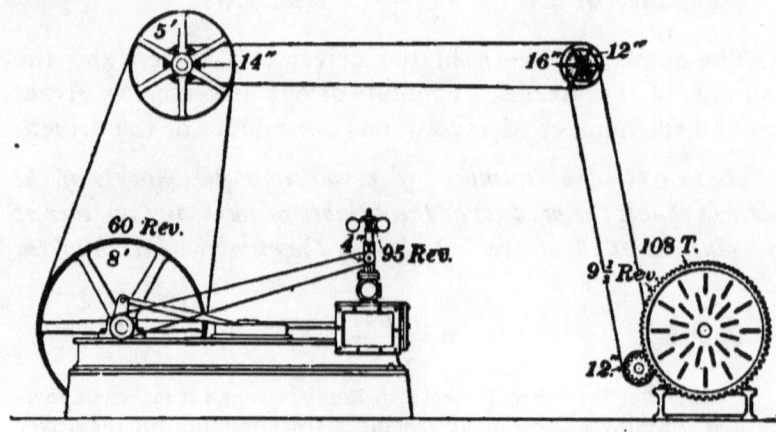

FIG. 90.

SOLUTION.—First determine the number of revolutions of the 4' pulley in order that the governor shall turn 95 times per minute. Applying rule **92**, $N = \dfrac{tn}{T} = \dfrac{30 \times 95}{19} = 150$ revolutions of gear on pulley shaft = revolutions of governor pulley. Now, applying rule **81**, the diameter of the pulley on the crank-shaft $= \dfrac{dn}{N} = \dfrac{4 \times 150}{60} = 10'$. Ans.

EXAMPLE.—In Fig. 90, the fly-wheel is 8 feet in diameter and drives a 5-foot pulley on the main shaft. A 14' pulley on the main shaft drives a 16' pulley on the countershaft. A 12' pulley on the countershaft drives a 12' pulley on a shaft on which is a pinion that meshes into a large gear attached to the face plate of a large lathe, and which has 108 teeth. How many teeth must the pinion have in order that the face plate may make $9\frac{1}{4}$ revolutions per minute?

SOLUTION.—Applying rule **83**, to find the revolutions per minute of the main shaft, $\dfrac{8 \times 60}{5} = 96$ R. P. M. Applying the same rule again to find the revolutions of the countershaft, $\dfrac{14 \times 96}{16} = 84$ R. P. M. Applying it once more to find revolutions of the pulley which turns the small gear, $\dfrac{12 \times 84}{12} = 84$ R. P. M. Applying rule **89**, $\dfrac{108 \times 9\frac{1}{4}}{84} = 12$ teeth in pinion or driver. Ans.

MECHANICS.

EXAMPLES FOR PRACTICE.

1. The driving pulley makes 110 R. P. M. and is 21" in diameter; what should be the size of the driven in order to make 385 R. P. M.?
Ans. 6".

2. The main shaft of a certain shop makes 120 R. P. M. It is desired to have the countershaft make 150 R. P. M. There are on hand pulleys of 16", 24", 28", 35", and 38" in diameter. Can two of these be used, or must a new pulley be ordered?
Ans. Use the 28" and 35" pulleys.

3. The pinion (driver) makes 174 R. P. M., and follower makes 24 R. P. M.; how many teeth must the pinion have if the follower has 87 teeth?
Ans. 12 teeth.

4. If an engine fly-wheel is 66" in diameter, and makes 160 R. P. M., what must be the diameter of the pulley on the main shaft to make 128 R. P. M.?
Ans. $82\frac{1}{2}$".

5. What is the pitch diameter of a gear whose pitch is $1\frac{1}{4}$", and has 28 teeth?
Ans. 11.14".

6. How many teeth are there in a gear whose pitch is .7854", and which is 23" in diameter?
Ans. 92 teeth.

7. What is the pitch of a gear whose diameter is 20.372", and which has 128 teeth?
Ans. $\frac{1}{2}$".

8. In a train of gears the drivers have 16, 30, 24, and 18 teeth, respectively; the followers have 12, 24, 36, and 40 teeth, respectively. If the first driver makes 80 R. P. M., how many R. P. M. will the last follower make?
Ans. 40 R. P. M.

FIXED AND MOVABLE PULLEYS.

512. Pulleys are also used for hoisting or raising loads, in which case the frame which supports the axle of the pulley is called the **block**.

513. A **fixed pulley** is one whose block is not movable, as in Fig. 91. In this case if the weight W be lifted by pulling down P, the other end of the cord W will evidently move the same distance upwards that P moves downwards; hence, P must equal W.

514. A **movable pulley** is one whose block is movable, as in Fig. 92. One end of the cord is fastened to the beam, and the weight is suspended

FIG. 91.

from the pulley, the other end of the cord being drawn up by the application of a force P. A little consideration will show that if P moves through a certain distance, say 1 foot, W will move through *half* that distance, or 6 inches; hence, a pull of one pound at P will lift 2 pounds at W.

The same would also be true if the free end of the cord were passed over a *fixed pulley*, as in Fig. 93, in which case the fixed pulley merely changes the direction in which P acts, so that a weight of 1 pound hung on the free end of the cord will balance 2 pounds hung from the *movable pulley*.

Fig. 92.

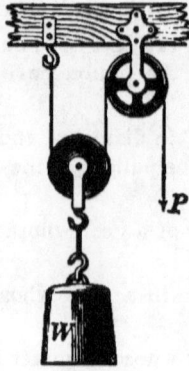

Fig. 93.

515. A combination of pulleys, as shown in Fig. 94, is sometimes used. In this case there are three movable and three fixed pulleys, and the amount of movement of W, owing to a certain movement of P, is readily found.

It will be noticed that there are *six parts* of the rope, not counting the free end; hence, if the movable block be lifted 1 foot, P remaining in the same position, there will be 1 foot of slack in each of the six parts of the rope, or *six feet* in all. Therefore, P would have to move 6 feet in order to take up this slack, or P moves 6 times as far as W. Hence, 1 pound at P will support 6 pounds at W, since the *power multiplied by the distance through which it moves equals the weight multiplied by the distance through which it moves.* It will also be noticed that there are three movable pulleys, and that $3 \times 2 = 6$.

Fig. 94.

516. Law of Combination of Pulleys:

Rule 93.—*In any combination of pulleys where one continuous rope is used, a load on the free end will balance a weight on the movable block as many times as great as itself as there are parts of the rope supporting the load, not counting the free end.*

The above law is good, whether the pulleys are side by side, as in the ordinary block and tackle, or whether they are arranged as in the figure.

EXAMPLE.—In a block and tackle having five movable pulleys, how great a force must be applied to the free end of the rope to raise 1,250 pounds?

SOLUTION.—Since there are five movable pulleys, there must be 10 parts of the rope to support them. Hence, according to the above law, a force applied to the free end will support a load 10 times as great as itself, or the force $= \frac{1,250}{10} = 125$ lb. Ans.

THE INCLINED PLANE.

517. An **inclined plane** is a slope, or a flat surface, making an angle with a horizontal line.

Three cases may arise in practice with the inclined plane:

1. Where the power acts parallel to the plane, as in Fig. 95.

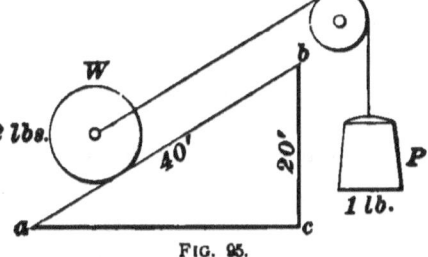

FIG. 95.

2. Where the power acts parallel to the base, as in Fig. 96.

3. Where the power acts at an angle to the plane or to the base, as in Fig. 97.

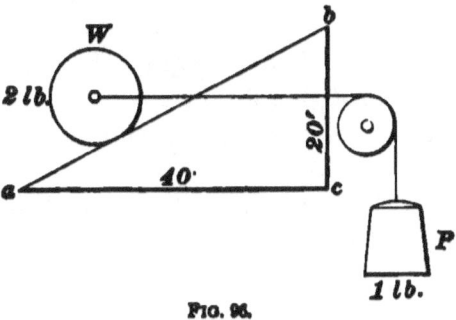

FIG. 96.

518. In Fig. 95 the relation existing between the power and the weight is easily found. The weight ascends a distance equal to cb, or the height of the inclined plane, while the power descends through a distance equal to ab, or the length of the inclined plane. Therefore,

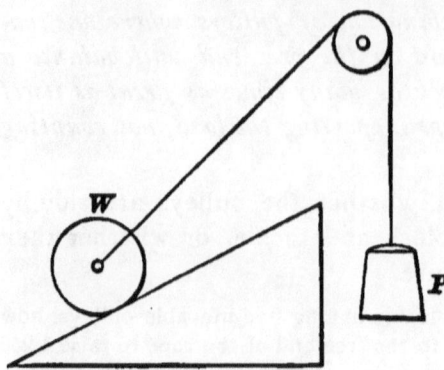

FIG. 97.

Rule 94.—*The power multiplied by the length of the inclined plane equals the weight multiplied by the height of the inclined plane;* or, letting

$$\text{length of plane} = l;$$
$$\text{base of plane} = b;$$
$$\text{height of plane} = h;$$

we have

$$P = \frac{Wh}{l} \text{ and } W = \frac{Pl}{h}.$$

EXAMPLE.—The length of an inclined plane is 40 feet, and its height is 5 feet; what force P will sustain a weight W of 100 lb.?

SOLUTION.—$P = \dfrac{Wh}{l} = \dfrac{100 \times 5}{40} = 12\frac{1}{4}$ lb. Ans.

EXAMPLE.—The length of an inclined plane is 8 feet, and its height is 15 inches; what weight will 120 lb. support?

SOLUTION.—$W = \dfrac{Pl}{h} = \dfrac{120 \times 96}{15} = 768$ lb. Ans.

In Fig. 96 the power is supposed to act parallel to the base for any position of W; therefore, while W is moving from the level ac to b, or through the height cb of the inclined plane, P will move through a distance equal to the length of the base ac. Hence, when the power acts parallel to the base, we have

Rule 95.—*The power multiplied by the base of the inclined plane equals the weight multiplied by the height of the plane,*

whence, $P = \dfrac{Wh}{b}$ and $W = \dfrac{Pb}{h}$.

MECHANICS. 187

EXAMPLE.—With a base 30 feet long, and a height of 6 feet, what power will sustain a weight of 75 lb.?

SOLUTION.—$P = \dfrac{75 \times 6}{30} = 15$ lb. Ans.

EXAMPLE.—A force of 12 lb. sustains a weight on a plane whose base is 6 ft. long, and height 18 inches. Find the weight.

SOLUTION.—$W = \dfrac{12 \times 6}{1\frac{1}{2}} = 48$ lb. Ans.

For Fig. 97 no rule can be given. The ratio of the power to the weight must be determined by trigonometry for every position of W; for, as W shifts its position, the angle that its cord makes with the face of the plane varies, and the magnitude of the force P depends on this angle; the smaller the angle the less is P required to be to support a given weight on a given plane.

519. The **wedge** is a movable inclined plane, and is used for moving a great weight a short distance. A common method of moving a heavy body is shown in Fig. 98.

Simultaneous blows of equal force are struck on the heads of the wedges, thus raising the weight W. The laws for wedges are the same as for Case 2 of the inclined plane.

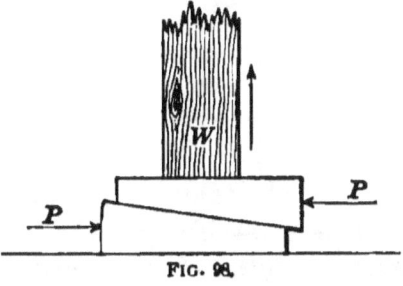

FIG. 98.

THE SCREW.

520. A **screw** is a cylinder with a helical projection winding around its circumference. This helix is called the *thread* of the screw. The distance that a point on the helix is drawn back or advanced in the direction of the length of the screw during one turn, is called the *pitch* of the screw.

MECHANICS.

The screw in Fig. 99 is turned in a *nut a*, by means of a force applied at the end of the handle *P*. For one complete revolution of the handle, the screw will be advanced lengthwise, *an amount equal to the pitch*. If the nut be fixed, and a weight be placed upon the end of the screw, as shown, it will be raised vertically a distance equal to the pitch by one revolution of the screw. During this revolution, the force at *P* will move through a distance equal to the circumference whose radius is *P F*. Therefore, $W \times$ pitch of thread $= P \times$ circumference of *P*.

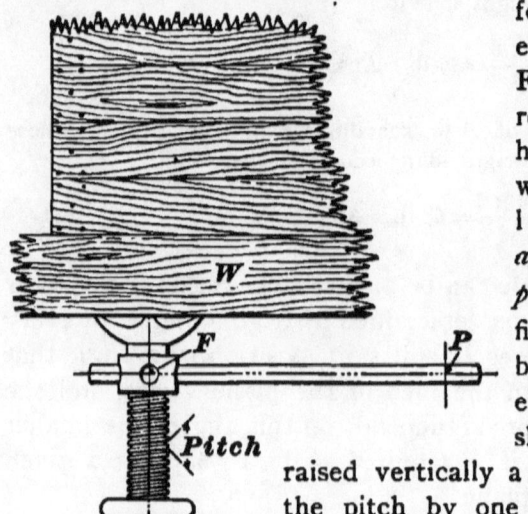

FIG. 99.

521. Hence, to find the weight that a given force will lift, we have

Rule 96.—*Multiply the force by its distance from the axis, or center of screw, and by 6.2832; divide this product by the pitch, and the quotient will be the weight required;*

or
$$W = \frac{P \times 6.2832 \times r}{p},$$

whence,
$$P = \frac{Wp}{6.2832 \times r}.$$

EXAMPLE.—It is desired to raise a weight by means of a screw having 5 threads per inch. The force is 40 pounds, and is applied at a distance of 14 inches from the center of the screw (*FP*, in Fig. 99); how great a weight can be raised?

SOLUTION.—Applying the formula,
$$W = \frac{40 \times 6.2832 \times 14}{\tfrac{1}{5}} = 17593 \text{ lb., very nearly. Ans.}$$

EXAMPLE.—The weight to be lifted is 4,000 pounds; the pitch of the screw is ¼ (that is, there are 4 threads per inch). The force is applied at a distance of 12 inches from the center of screw; how great must it be?

SOLUTION.—Applying the formula,

$$P = \frac{4{,}000 \times \frac{1}{4}}{6.2832 \times 12} = \text{rather more than } 13\frac{1}{4} \text{ lb., nearly. Ans.}$$

522. Single-threaded screws of less than 1 inch pitch are generally classified by the number of threads they have in 1 inch of their length. In such cases *one inch divided by the number of threads equals the pitch;* thus, the pitch of a screw that has 8 threads per inch is $\frac{1}{8}$; one of 32 threads per inch is $\frac{1}{32}$, etc.

523. Velocity Ratio.—The ratio of the distance through which the power moves to the corresponding distance through which the weight moves is called the **velocity ratio.**

Thus, if the power move through 12 inches while the weight moves through 1 inch, the velocity ratio is 12 to 1, or 12; that is, P moves 12 times as fast as W.

If the velocity ratio be known, the weight which any machine will raise can be found by multiplying the power *by the velocity ratio.* If the velocity ratio is 8.7 to 1, or 8.7, $W = 8.7 \times P$, since $W \times 1 = P \times 8.7$.

NOTE.—In all of the preceding cases, including the last, the effect of friction has been neglected.

FRICTION.

524. Friction is the resistance that a body meets with from the surface on which it moves.

525. The **ratio** between the *resistance* to the motion of a body due to friction and the *perpendicular* pressure between the surfaces is called the **coefficient of friction.**

If a weight W, as in Fig. 100, rests upon a horizontal plane, and has a cord fastened to it passing over a pulley a, from which a weight P is suspended, then, if P is just sufficient to start W, the ratio of P to W, or $\frac{P}{W}$, is the *coefficient of friction* between W and the surface upon which it slides.

The weight W is the perpendicular pressure, and P is the

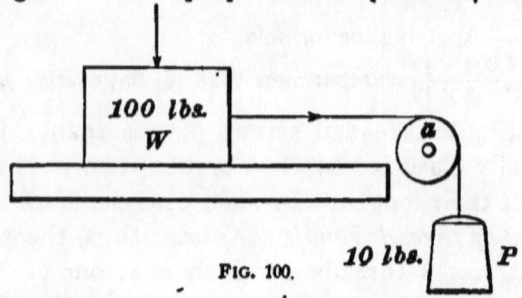

FIG. 100.

force necessary to overcome the resistance to the motion of W, due to friction.

If $W = 100$ pounds and $P = 10$ pounds, the coefficient of friction for this particular case would be $\dfrac{P}{W} = \dfrac{10}{100} = .1$.

526. Laws of Friction:

1. Friction is directly proportional to the perpendicular pressure between the two surfaces in contact.

2. Friction is independent of the extent of the surfaces in contact when the total perpendicular pressure remains the same.

3. Friction increases with the roughness of the surfaces.

4. Friction is greater between surfaces of the same material than between those of different materials.

5. Friction is greatest at the beginning of motion.

6. Friction is greater between soft bodies than between hard ones.

7. Rolling friction is less than sliding friction.

8. Friction is diminished by polishing or lubricating the surfaces.

527. Law 1 shows why the friction is so much greater on journals after they begin to heat than before. The heat causes the journal to expand, thus increasing the pressure between the journal and its bearing, and, consequently, increasing the friction.

MECHANICS.

Law 2 states that, no matter how small the surface may be which presses against another, if the perpendicular pressure is the same the friction will be the same. Therefore, large surfaces are used where possible, not to reduce the friction, but to reduce the wear and diminish the liability of heating.

528. For instance, if the perpendicular pressure between a journal and its bearing is 10,000 pounds, and the coefficient of friction is .2, the amount of friction is $10,000 \times .2 = 2,000$ pounds.

Suppose that one-half the area of the surface of the journal is 80 square inches, then, the amount of friction for each square inch of bearing is $2,000 \div 80 = 25$ pounds.

If half the area of the surface had been 160 square inches, the friction would have been the same, that is, 2,000 pounds; but the friction per square inch would have been $2,000 \div 160 = 12\frac{1}{2}$ pounds, just one-half as much as before, and the wear and liability to heat would be one-half as great also.

TABLE OF COEFFICIENTS OF FRICTION.
TABLE 16.

Description of Surfaces in Contact.	Disposition of Fibers.	State of the Surfaces.	Coefficient of Friction.
Oak on oak.	Parallel	Dry	.48
Oak on oak.	Parallel	Soaped	.16
Wrought iron on oak.	Parallel	Dry	.62
Wrought iron on oak.	Parallel	Soaped	.21
Cast iron on oak.	Parallel	Dry	.49
Cast iron on oak.	Parallel	Soaped	.19
Wrought iron on cast iron.	—	Slightly Unctuous	.18
Wrought iron on bronze.	—	Slightly Unctuous	.18
Cast iron on cast iron.	—	Slightly Unctuous	.15

529. The power which is required to raise a weight, or overcome an equal resistance in any machine, is thus always *greater than this weight or resistance divided by the velocity ratio of the machine.*

Thus, if there were no friction, a machine whose velocity ratio was 5 would, by an application of 100 pounds of force, raise a weight of 500 pounds.

Now, suppose that the friction in the machine is equivalent to 10 pounds of force; then, it would take 110 pounds of force to raise 500 pounds.

If, in the above illustration, friction were neglected, 110 pounds $\times 5 = 550$ pounds, or the weight that 110 pounds would raise; but, owing to the frictional resistance, it only raised 500 pounds. Therefore, we have for the ratio between the two, $\dfrac{500}{550} = .91$. That is, $500 : 550 :: .91 : 1$.

530. Efficiency.—This ratio between the weight actually raised and the power multiplied by the velocity ratio, is called the **efficiency of the machine.**

For example, if the weight actually raised by a machine, say a screw, is 1,600 pounds, and the power multiplied by the velocity ratio is 2,400 pounds, the efficiency of this machine is

$$\frac{1,600}{2,400} = .66\tfrac{2}{3}, \text{ or } 66\tfrac{2}{3}\%.$$

EXAMPLE.—In a machine having a combination of pulleys and gears, the velocity ratio of the whole is 9.75. A force of 250 pounds just lifts a weight of 1,626 pounds. What is the efficiency of the machine?

SOLUTION.—Efficiency $= \dfrac{1,626}{250 \times 9.75} = .6671$, or 66.71%. Ans.

Since the total amount of friction varies with the load, it follows that the efficiency will also vary for different loads.

531. If the efficiency of a machine is known, the force actually required to raise a given load may be found by dividing the load by the product of the velocity ratio of the machine and the efficiency. Thus, if a certain machine has a velocity ratio of 10.6, and its efficiency is 60%, the force which must actually be applied to raise a load of 840 pounds is $840 \div 10.6 \times .60 = 840 \div 6.36 = 132.1$ pounds, nearly. If there had been no losses through friction, etc., the force required would have been $840 \div 10.6 = 79.25$ pounds, nearly.

If the efficiency is known, the weight which a certain force will raise may be found by multiplying together the force, velocity ratio, and the efficiency. Thus, if a certain machine has a velocity ratio of $6\frac{1}{2}$, and an efficiency of 78%, a force of 140 pounds will raise a weight of $140 \times 6\frac{1}{2} \times .78 = 709.8$ pounds.

532. When finding the force necessary to overcome the friction, the *perpendicular pressure* on the surface considered must always be taken. In order to find (approximately) the maximum force that is required to overcome the friction between cross-head and guides, we have

Rule 97.—*Multiply the total piston pressure by the length of crank and by the coefficient of friction and divide by the length of main rod;*

or, letting $P =$ total piston pressure;
$l =$ length of main rod;
$r =$ length of crank;
$c =$ coefficient of friction between cross-head and guides;
$F =$ force required to overcome this friction,

we have $$F = \frac{Prc}{l}.$$

EXAMPLE.—An engine whose piston is 16″ in diameter carries a steam pressure of 80 lb. per sq. in. If the crank is 12″ long, and the connecting-rod is 66″ long, what is the perpendicular pressure on the guides? The coefficient of friction for this case being 12%, what force will be required to overcome the friction?

SOLUTION. — Pressure on piston $= 16^2 \times .7854 \times 80 = 16{,}085$ lb. $\frac{16{,}085 \times 12}{66} = 2{,}924.55$ lb. $=$ perpendicular pressure. Ans. $2{,}924.55 \times .12 = 350.95$ lb. $=$ force required to overcome the friction. Ans.

MECHANICS.

EXAMPLES FOR PRACTICE.

1. How great a force must be applied to the free end of the rope of a block and tackle which has four movable pulleys, to raise a weight of 746 lb.? Ans. 93¼ lb.

2. An inclined plane is 30 ft. long and 7 ft. high; what force is required to roll a barrel of flour weighing 196 lb. up the plane, friction being neglected? Ans. 45.7¼ lb.

3. The distance from the axis of a screw to the point on the handle where the force is applied is 12'. The screw has 8 threads per inch. What force is necessary to raise a weight of 1,248 lb.?
Ans. 2.07 lb., nearly.

4. In example 3, what should be the length of the handle to raise a weight of 5,400 lb. by the application of a force of 20 lb.?
Ans. 5.371', nearly.

5. What is the velocity ratio (a) in example 3? (b) In example 4?
Ans. $\begin{cases} (a)\ 603,\ \text{nearly.} \\ (b)\ 270. \end{cases}$

6. An engine piston is 24' in diameter. If the steam pressure is 93 lb. per sq. in.; the length of the connecting-rod, 8 ft. 4"; the length of crank, 20", and coefficient of friction, 14%, what is (a) the greatest perpendicular pressure on the guides? (b) The force required to overcome the friction?
Ans. $\begin{cases} (a)\ 8,414.46\ \text{lb.} \\ (b)\ 1,178\ \text{lb.} \end{cases}$

CENTRIFUGAL FORCE.

533. If a body be fastened to a string and whirled, so as to give it a circular motion, there will be a pull on the string which will be greater or less, according as the velocity increases or decreases. The cause of this pull on the string will now be explained.

Suppose that the body is revolved horizontally, so that the action of gravity upon it will always be the same. According to the first law of motion, a body put in motion tends to move in a straight line, unless acted upon by some other force, causing a change in the direction. When the body moves in a circle, the force that causes it to move in a circle instead of a straight line is exactly equal to the tension of the string. If the string were cut, the pulling

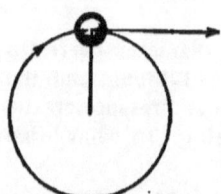
FIG. 101.

force that drew it away from the straight line would be removed, and the body would then "fly off at a tangent;" that is, it would move in a straight line tangent to the circle, as shown in Fig. 101.

Since, according to the third law of motion, every action has an equal and opposite reaction, we call that force which acts as an equal and opposite force to the pull of the string the **centrifugal force**, and it acts *away* from the center of motion.

534. The other force, or tension, of the string is called the **centripetal force**, and it acts *towards* the center of motion. It is evident that these two forces, acting in opposite directions, tend to pull the string apart, and, if the velocity be increased sufficiently, the string will break. It is also evident that no body can revolve without generating centrifugal force.

535. The value of the centrifugal force, expressed in pounds, of any revolving body is calculated by the following rule:

Rule 98.—*The centrifugal force equals the continued product of .00034, the weight of the body in pounds, the radius in feet (taken as the distance between the center of gravity of the body and the center about which it revolves), and the square of the number of revolutions per minute.*

Let F = centrifugal force in pounds;
   W = weight of revolving body in pounds;
   R = radius in feet of circle described by center of gravity of revolving body;
   N = revolutions per minute of revolving body; then
   $F = .00034\ W R N^2.$

536. In calculating the centrifugal force of fly-wheels, it is the usual practice to consider the rim of the wheel only, and not take the arms and hub of the wheel into account. In this case R would be taken as the *distance between the center of the rim and the center of the shaft.*

MECHANICS.

EXAMPLE.—What would be the centrifugal force developed by a cast-iron fly-wheel, whose outside diameter was 10 feet, width of face 20 inches, and thickness of rim 6 inches, turning at the rate of 80 revolutions per minute? Take the weight of a cubic inch of cast iron as .261 lb.

SOLUTION.—First calculate the weight of the rim. The diameter of the rim = $10 \times 12 = 120$ inches; the diameter of the circle midway between the inside and outside diameters of the rim = $120 - 6 = 114$ inches. The number of cubic inches in the rim = $114 \times 3.1416 \times 20 \times 6 = 42,977$ cubic inches. $42,977 \times .261 = 11,217$ pounds = weight.

Radius = $\frac{114}{2} \div 12 = 4\frac{3}{4}$ feet. R. P. M. = 80.

Hence, by rule **98**, centrifugal force = $.00034 \times 11,217 \times 4\frac{3}{4} \times 80^2 = 115,939$ pounds. Ans.

SPECIFIC GRAVITY.

537. The **specific gravity of a body** *is the ratio between its weight and the weight of a like volume of water.*

538. Since gases are so much lighter than water, it is usual to take the specific gravity of a gas as the ratio between the weight of a certain volume of the gas and the weight of the same volume of air.

EXAMPLE.—A cubic foot of cast iron weighs 450 pounds; what is its specific gravity, a cubic foot of water weighing 62.5 pounds?

SOLUTION.— $\frac{450}{62.5} = 7.2$. Ans.

539. The specific gravities of different bodies are given in printed tables; hence, if it is desired to know the weight of a body that can not be conveniently weighed, *calculate its cubical contents. Multiply the specific gravity of the body by the weight of a like volume of water, remembering that a cubic foot of water weighs 62.5 pounds.*

EXAMPLE.—How much will 3,214 cubic inches of cast iron weigh? Take its specific gravity as 7.21.

SOLUTION.—Since 1 cubic foot of water weighs 62.5 pounds, 3,214 cubic inches weigh

$\frac{3,214}{1,728} \times 62.5 = 116.25$ pounds.

$116.25 \times 7.21 = 838.16$ pounds. Ans.

MECHANICS. 197

EXAMPLE.—What is the weight of a cubic inch of cast iron?

SOLUTION.— $\dfrac{62.5}{1,728} \times 7.21 = .2608$ pound. Ans.

NOTE.—One cubic foot of pure distilled water at a temperature of 39.2° Fahrenheit weighs 62.42 pounds, but the value usually taken in making calculations is 62½ pounds.

EXAMPLE.—What is the weight in pounds of 7 cubic feet of oxygen?

SOLUTION.—One cubic foot of air weighs .08073 lb., and the specific gravity of oxygen is 1.1056 compared with air; hence,

$.08073 \times 1.1056 \times 7 = .62479$ pound, nearly.

EXAMPLES FOR PRACTICE.

1. The balls of a steam-engine governor each weigh 5 pounds. If they revolve in a circle whose diameter is 14" at the rate of 80 revolutions per minute, what is the centrifugal force of each ball?
Ans. 6.347 lb., nearly.

2. The rim of a cast-iron engine fly-wheel has an outside diameter of 15 feet. If the rim is 8" thick and 12" wide, and the fly-wheel makes 40 R. P. M., what is the centrifugal force of the rim? Ans. 52,785 lb.

3. If a cubic foot of a certain alloy weighs 678 pounds, what is its specific gravity? Ans. 10.848.

4. What is the weight of (*a*) 12.4 cubic inches of lead? (*b*) of steel? (*c*) of aluminum?
Ans. $\begin{cases}(a)\ 5.0964\text{ lb.}\\(b)\ 3.5216\text{ lb.}\\(c)\ 1.116\text{ lb.}\end{cases}$

5. The specific gravity of an alloy of lead and zinc is 8.26; what is the weight of a cubic foot? Ans. 516.25 lb.

WORK.

540. Work *is the overcoming of resistance continually occurring along the path of motion.*

Mere motion is not work, but if a body in motion constantly overcomes a resistance, it does work.

541. *The* **measure of work** *is one pound raised vertically one foot,* and is called **one foot-pound.** All work is measured by this standard. A horse going up hill does an amount of work equal to his own weight plus the weight of the wagon and contents plus the frictional resistances reduced to an equivalent weight multiplied by the vertical height of the hill. Thus, if the horse weighs 1,200 pounds, the wagon and contents 1,200 pounds, and the frictional resistances equal 400 pounds, then, if the vertical height of the hill is 100 feet, the work done is equal to $(1{,}200 + 1{,}200 + 400) \times 100 = 280{,}000$ foot-pounds.

Rule 99.—*In all cases the force (or resistance) multiplied by the distance through which it acts equals the work. If a weight be raised, the weight multiplied by the vertical height of the lift equals the work.*

542. The *total* amount of work is independent of time, whether it takes one minute or one year in which to do it; but in order to compare the work done by different machines with a common standard, time must be considered. If one machine does a certain amount of work in 10 minutes, and another machine does exactly the same amount of work in 5 minutes, the second machine can do twice as much work as the first in the same time.

543. The common standard to which all work is reduced is the **horsepower.**

One horsepower *is 33,000 foot-pounds per minute; in other words, it is 33,000 pounds raised vertically one foot in one minute, or 1 pound raised vertically 33,000 feet in one minute, or any combination that will, when multiplied together, give 33,000 foot-pounds in one minute.*

544. Thus, the work done in raising 110 pounds vertically 5 feet in one second is a horsepower; for, since in one minute there are 60 seconds, $110 \times 5 \times 60 = 33{,}000$ foot-pounds in one minute.

EXAMPLE.—If the coefficient of friction is .3, how many horsepower will it require to draw a load of 10,000 pounds on a level surface, a distance of one mile in one hour?

SOLUTION.—$10{,}000 \times .3 = 3{,}000$ pounds = the force necessary to overcome the resistance (resistance of the air is neglected). One mile = 5,280 feet; one hour = 60 minutes. Therefore, $\frac{5{,}280}{60} = 88$ feet per minute.

Work done = force multiplied by the space = $3{,}000 \times 88 = 264{,}000$ foot-pounds per minute.

Horsepower $= \frac{264{,}000}{33{,}000} = 8$ horsepower. Ans.

The abbreviation for horsepower is H. P.

MECHANICS.

545. Energy is a term used to express *the ability of an agent to do work.* Work can not be done without motion, and the work that a moving body is capable of doing in being brought to rest is called the **kinetic energy** of the body.

Kinetic energy means the actual visible energy of a body in motion. The work which a moving body is capable of doing in being brought to rest is exactly the same as the kinetic energy developed by it in falling in a vacuum through a height sufficient to give it the same velocity.

Rule 100.—*The kinetic energy of a moving body in foot-pounds equals its weight in pounds multiplied by the square of its velocity in feet per second, and divided by 64.32.*

If W represents the weight of the body in pounds, and v the velocity in feet per second,

$$\text{kinetic energy} = \frac{Wv^2}{64.32}.$$

If a weight is raised a certain height, a certain amount of work is done equal to the product of the weight and the vertical height. If a weight is suspended at a certain height and allowed to fall, it will do the same amount of work in foot-pounds that was required to raise the weight to the height through which it fell.

EXAMPLE.—If a body weighing 25 pounds falls from a height of 100 feet, how much work can it do?

SOLUTION.—Work $= Wh = 25 \times 100 = 2{,}500$ foot-pounds. Ans.

546. It requires the same amount of work or energy to stop a body in motion within a certain time as it does to give it that velocity in the same time.

EXAMPLE.—A body weighing 50 pounds has a velocity of 100 feet per second; what is its kinetic energy?

SOLUTION.—Kinetic energy $= \dfrac{Wv^2}{64.32} = \dfrac{50 \times 100^2}{64.32} = 7{,}773.63$ foot-pounds. Ans.

EXAMPLE.—In the last example, how many horsepower will be required to give the body this amount of kinetic energy in 3 seconds?

SOLUTION.—1 H. P. $= 33{,}000$ pounds raised 1 foot in one minute.

If 7,773.63 foot-pounds of work are done in 3 seconds, in one second there would be done $\frac{7,773.65}{3} = 2,591.21$ foot-pounds of work. One horsepower = 33,000 ft.-lb. per min. = 33,000 ÷ 60 = 550 ft.-lb. per sec.

The number of horsepower developed will be

$$\frac{2,591.21}{550} = 4.7113 \text{ H. P.} \quad Ans.$$

547. Potential energy *is latent energy; it is the energy which a body at rest is capable of giving out under certain conditions.*

If a stone is suspended by a string from a high tower, it has potential energy. If the string is cut, the stone will fall to the ground, and during its fall its potential energy will change into kinetic energy, so that at the instant it strikes the ground its potential energy is wholly changed into kinetic energy.

At a point equal to one-half the height of the fall, the potential and kinetic energies are equal. At the end of the first quarter the potential energy was $\frac{3}{4}$, and the kinetic energy $\frac{1}{4}$; at the end of the third quarter the potential energy was $\frac{1}{4}$, and the kinetic energy $\frac{3}{4}$.

A pound of coal has a certain amount of potential energy. When the coal is burned, the potential energy is liberated and changed into kinetic energy in the form of heat. The kinetic energy of the heat changes water into steam, which thus has a certain amount of potential energy. The steam acting on the piston of an engine causes it to move through a certain space, thus overcoming a resistance, changing the potential energy of the steam into kinetic energy, and thus doing work.

Potential energy *then, is the energy stored within a body which may be liberated and produce motion, thus generating kinetic energy, and enabling work to be done.*

548. The principle of **conservation of energy** teaches that energy, like matter, can never be destroyed. If a clock is put in motion, the potential energy of the spring is changed into kinetic energy of motion, which turns the wheels, thus producing friction.

The friction produces heat, which dissipates into the surrounding air, but still the energy is not destroyed—it merely exists in another form. The potential energy in coal was received from the sun, in the form of heat, ages ago, and has lain dormant for millions of years.

BELTS.

549. A **belt** is a flexible connecting band which drives a pulley by its frictional resistance to slipping at the surface of the pulley. Belts are most commonly made of leather or rubber, and united in long lengths by *cementing, riveting,* or *lacing.*

550. Belts are made *single* and *double.* A **single belt** is one composed of a single thickness of leather; a **double belt** is one composed of two thicknesses of leather cemented and riveted together the whole length of the belt.

551. To find the length of a belt:

In practice, the necessary length for a belt to pass around pulleys that are already in their position on a shaft is usually obtained by passing a tape line around the pulleys, the stretch of the tape line being allowed as that necessary for the belt. The lengths of open-running belts for pulleys not in position can be obtained as follows:

Rule 101.—*The length of a belt for open-running pulleys equals $3\frac{1}{4}$ times one-half the sum of the diameters of the pulleys plus 2 times the distance between the centers of the shaft.*

Let D = diameter of one pulley;
D_1 = diameter of other pulley;
L = distance between the centers of the shafts;
B = length of the belt.

Then, $$B = 3\tfrac{1}{4}\left(\frac{D+D_1}{2}\right) + 2L.$$

EXAMPLE.—The distance between the centers of two shafts is 9 feet 7 inches; the diameter of the large pulley is 36 inches, and the diameter of the small one is 14 inches; what is the necessary length of the belt?

SOLUTION.—By formula, $B = 3\tfrac{1}{4}\left(\frac{D+D_1}{2}\right) + 2L.$

Substituting the values given, we have, since 9 ft. 7' = 115',

$$B = 3\tfrac{1}{4}\left(\frac{36+14}{2}\right) + 2 \times 115 = 311\tfrac{1}{4} \text{ in., or 25 ft. } 11\tfrac{1}{4} \text{ in. Ans.}$$

552. To find the width of a single leather belt that will transmit any given horsepower when equal pulleys are used:

Rule 102.—*The width of the belt in inches equals 800 times the horsepower to be transmitted divided by the speed of the belt in feet per minute.*

Let W = width of belt in inches;

H = horsepower to be transmitted;

S = speed of belt in feet per minute; then,

$$W = \frac{800\,H}{S}.$$

EXAMPLE.—What width of single leather belt is required to transmit 20 horsepower when equal pulleys are used, and the speed is 1,600 feet per minute?

SOLUTION.—Formula, $W = \dfrac{800\,H}{S}.$

Substituting, we have

$$W = \frac{800 \times 20}{1,600} = 10 \text{ inches. Ans.}$$

553. To find the number of horsepower that a single leather belt will transmit, its width and speed being given:

Rule 103.—*The number of horsepower equals the product of the width in inches and the speed in feet per minute divided by 800.*

Or, $$H = \frac{WS}{800}.$$

EXAMPLE.—If a 10′ single leather belt is to be run at a speed of 1,600 feet per minute, what horsepower will it transmit?

SOLUTION.—Formula, $H = \dfrac{WS}{800}.$

Substituting, we have

$$H = \frac{10 \times 1,600}{800} = 20 \text{ horsepower. Ans.}$$

554. When the pulleys are of different diameters, the arc of contact must be considered. To find the number of degrees in the arc of contact, *multiply the length of belt in contact on the smaller pulley by 360, and divide the product by the circumference of the pulley, calculating the result to the nearest whole number. The quotient is the arc of contact.*

Having found the arc of contact, *subtract it from 180° and multiply the result by 3. Add this last result to 800; the number thus obtained should be used instead of 800, in rules 102 and 103.*

EXAMPLE.—What should be the width of a single leather belt to transmit 25.24 horsepower at a speed of 1,500 feet per minute, the diameter of the smaller pulley being 24", and the belt having 30" of its length in contact with it?

SOLUTION.—Arc of contact $= \dfrac{30 \times 360}{24 \times 3.1416} = 143°$. $(180 - 143) \times 3 = 111$. $800 + 111 = 911$. Using rule **102**, and 911 instead of 800,

$$W = \dfrac{911 \times 25.24}{1,500} = 15.33", \text{ say } 15\tfrac{1}{3}". \text{ Ans.}$$

555. To find the width of a double belt that will transmit the same horsepower as a given single belt, let W_1 represent the width of the double belt; then,

Rule 104.—*Multiply the width of a single belt that will transmit the same horsepower by $\tfrac{2}{3}$.*

Or, $W_1 = \tfrac{2}{3} W$.

EXAMPLE.—If a single leather belt is 15" in width, and transmits 22 horsepower, what must be the width of a double belt to transmit the same horsepower?

SOLUTION.—Applying the rule, $15 \times \tfrac{2}{3} = 10$ in. $=$ width of double belt. Ans.

556. Lacing Belts.—Many good methods of fastening the ends of belts are employed, but lacing is generally used, as it is flexible, like the belt, and runs noiselessly over the pulleys.

When punching a belt for lacing, use an oval punch, the long diameter of the hole to be parallel with the side of the belt.

In a 3-inch belt there should be four holes in each end, two in each row. In a 6-inch belt, seven holes, four in the row nearest the end. A 10-inch belt should have nine holes, five in the row nearest the end. The edges of the holes should not be nearer than $\tfrac{3}{4}$ of an inch from the sides, and $\tfrac{1}{4}$ of an inch from the ends of the belt. The second row should be at least $1\tfrac{3}{4}$ inches from the end.

Always begin to lace from the center of the belt, and take

care to get the ends exactly in line. The lacing should not be crossed on the side of the belt that runs next to the pulley. Always run the hair side of the belt next to the pulley.

HORSEPOWER OF GEARS.

557. To find the horsepower which can be safely transmitted by gears whose face, or breadth of tooth, is from $2\frac{1}{2}$ to 3 times their pitch:

Rule 105.—*The horsepower which can be safely transmitted equals the continued product of the square of the pitch, the velocity in feet per minute and .01.*

Let $p =$ the pitch;

$s =$ circumferential speed of a point on the pitch circle in feet per minute; then,

$$H.\ P. = .01\ s\ p^2.$$

EXAMPLE.—What horsepower can be safely transmitted by a gear whose pitch diameter is 66.84 in., pitch $1\frac{3}{4}$ in., and which makes 60 R. P. M.?

SOLUTION.—The velocity which is to be used when applying rule **105** is the circumferential speed of a point on the pitch circle.

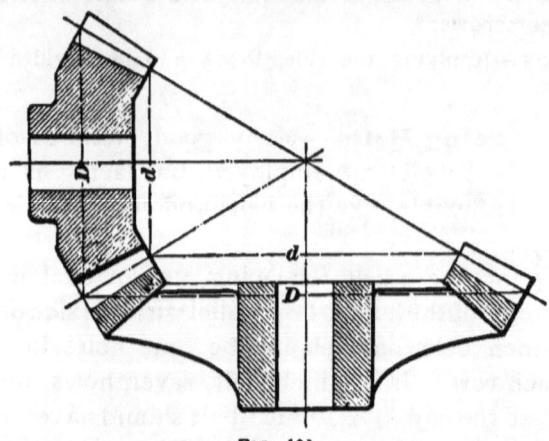

FIG. 102.

Hence, $66.84 \times 3.1416 = 209.98$ in. $=$ circumference of pitch circle $= \frac{209.98}{12}$ ft. $\frac{209.98}{12} \times 60 = 1,049.9 =$ velocity in ft. per min.

Now, applying formula, H. P. $= .01 \times 1,049.9 \times 1.75^2 = 32.1532$ horsepower. Ans.

558. When measuring bevel gears, the diameter of the largest pitch circle should be taken, as D, Fig. 102.

When calculating their horsepower, use the small, or inner, diameter, as d, Fig. 102. Either diameter may be used when calculating the revolutions per minute or number of teeth, by rules **86-92**, but if the inner or outer diameter of one gear be used, the corresponding diameter of the other gear which meshes with it must also be used.

EXAMPLES FOR PRACTICE.

1. How many foot-pounds of work are required to overcome for 7 minutes the friction of the cross-head of an engine which has a stroke of 4 ft., and makes 160 strokes per minute, if the coefficient of friction is 8%, and the average perpendicular pressure is 12,460 lb.?
Ans. 4,465,664 ft.-lb.

2. In the above example, what horsepower is required?
Ans. 19.332 H. P.

3. A cannon ball weighing 500 lb. is fired with a velocity of 1,600 ft. per sec.; what is its kinetic energy?   Ans. 19,900,497.5 ft.-lb.

4. An open belt drives two pulleys which are, respectively, 42" and 20" in diameter, and 23 ft. apart between their centers; what should be the length of the belt?   Ans. $652\frac{1}{4}$", or 54 ft. $4\frac{1}{4}$".

5. What width of single leather belting, which has 2 ft. 9" contact on the small pulley, is required to transmit 10 horsepower at a speed of 1,500 ft. per min.? Give width to nearest half inch. Diameter of small pulley 26".   Ans. 6".

6. What should be the width of the main belt of a steam engine to transmit 120 horsepower? The engine runs at 80 R. P. M., the band-wheel is 8 ft. in diameter, the belt is double, and has a contact of 6 ft. on the smaller pulley, which is 5 ft. in diameter. Take the speed of the belt the same as that of a point on the circumference of the band-wheel.   Ans. $36\frac{1}{4}$".

7. A 26" double belt runs at a speed of 2,830 ft. per min., and has a contact of 5 ft. on the smaller pulley; what horsepower is it transmitting? Diameter of small pulley is 48".   Ans. 121.15 H. P.

8. What horsepower can be safely transmitted by a gear whose pitch is $2\frac{1}{4}$", pitch diameter 44.66", and which makes 80 R. P. M.?
Ans. 42.24 H. P.

MECHANICS.
(CONTINUED.)

HYDROSTATICS.

559. Hydrostatics treats of liquids at rest under the action of forces.

560. Liquids are very nearly *incompressible*. A pressure of 15 pounds per square inch compresses water less than $\frac{1}{20000}$ of its volume.

561. Fig. 103 represents two cylindrical vessels of exactly the same size. The vessel a is fitted with a wooden block of the same size as the cylinder, and can move in it; the vessel b is filled with water, whose depth is the same as the length of the wooden block in a. Both vessels are fitted with air-tight pistons P, whose areas are each 10 sq. in.

Suppose, for convenience, that the weights of the pistons, block, and water be neglected, and that a force of 100 pounds be applied to both pistons. The pressure

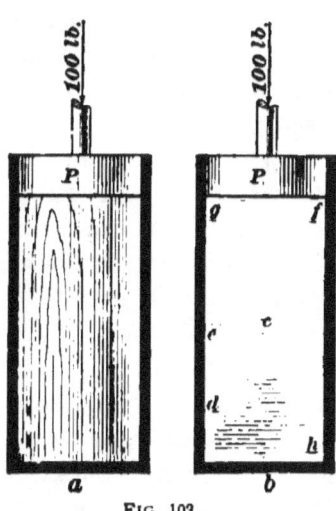

FIG. 103.

per square inch will be $\frac{100}{10} = 10$ pounds. In the vessel a this pressure will be transmitted to the bottom of the vessel, and will be 10 pounds per square inch; it is easy to see that there will be no pressure on the sides. In the vessel b an entirely different result is obtained. The pressure on the bottom will be the same as in the other case, that is, 10 pounds per square inch, but, owing to the fact that the molecules of the water are perfectly free to move, this pressure of 10 pounds per square inch is *transmitted in every direction with the same intensity;* that is to say, the pressure at any point c, d, e, f, g, h, etc., due to the force of 100 pounds, is exactly the same, and equals 10 pounds per square inch.

562. This may be easily proven experimentally by means of an apparatus like that shown in Fig. 104. Let the area of the piston a be 20 sq. in.; of b, 7 sq. in.; of c, 1 sq. in.; of d, 6 sq. in.; of e, 8 sq. in., and of f, 4 sq. in.

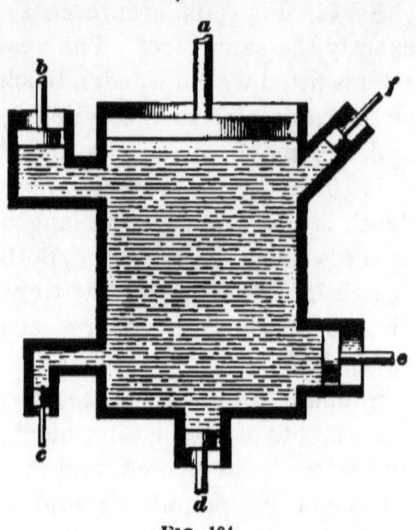

FIG. 104.

If the pressure due to the weight of the water be neglected, and a force of 5 pounds be applied at c (whose area is 1 sq. in.), a pressure of 5 pounds per square inch will be transmitted in all directions, and in order that there shall be no movement, a force of $6 \times 5 = 30$ pounds must be applied at d, 40 pounds at e, 20 pounds at f, 100 pounds at a, and 35 pounds at b.

If a force of 99 pounds were applied to a, instead of 100 pounds, the piston a would rise, and the other pistons b, c, d, e, and f, would move inwards; but, if the force applied to a were 100 pounds, they would all be in equilibrium. If 101 pounds were applied at a, the pressure per square inch would be $\frac{101}{20} = 5.05$ pounds, which would be transmitted in all directions; and, since the pressure due to the load on c is only 5 pounds per square inch, it is now evident that the piston a will move downwards, and the pistons b, c, d, e, and f will be forced outwards.

MECHANICS.

563. The whole of the preceding may be summed up as follows:

The pressure per unit of area exerted anywhere upon a mass of liquid is transmitted undiminished in all directions, and acts with the same force upon all surfaces, in a direction at **right angles** *to those surfaces.*

This law was first discovered by Pascal, and is the most important in **hydromechanics.** Its meaning should be thoroughly understood.

EXAMPLE.—If the area of the piston c, in Fig. 104, were 8.25 sq. in., and a force of 150 pounds were applied to it, what forces would have to be applied to the other pistons to keep the water in equilibrium, assuming that their areas were the same as given before?

SOLUTION.— $\frac{150}{8.25} = 18.182$ pounds per square inch, nearly.

$20 \times 18.182 = 363.64$ lb. = force to balance a.
$7 \times 18.182 = 127.274$ lb. = force to balance b.
$1 \times 18.182 = 18.182$ lb. = force to balance c. ⎬ Ans.
$6 \times 18.182 = 109.092$ lb. = force to balance d.
$4 \times 18.182 = 72.728$ lb. = force to balance f.

564. The pressure due to the weight of a liquid *may be downwards, upwards, or sideways.*

565. Downward Pressure.—In Fig. 105 the pressure on the bottom of the vessel a is, of course, equal to the weight of the water it contains. If the area of the bottom of the vessel b, and the depth of the liquid contained in it, are the same as in the vessel a, the pressure on the bottom of b will be the same as on the bottom of a. Suppose the bottoms of the vessels a and b are 6 inches square, and that the part $c\ d$ in the vessel b is 2 inches square, and that they are filled with water. Then, the weight of one cubic inch of water is $\frac{62.5}{1,728}$ pound =

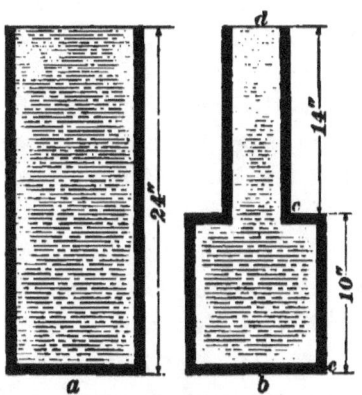

FIG. 105.

.03617 pound. The number of cubic inches in $a = 6 \times 6 \times 24 = 864$ cubic inches; and the weight of the water is $864 \times .03617 = 31.25$ pounds. Hence, the total pressure on the bottom of the vessel a is 31.25 pounds, or 0.868 pound per square inch.

The pressure in b, due to the weight contained in the part bc is $6 \times 6 \times 10 \times .03617 = 13.02$ pounds.

The weight of the part contained in cd is $2 \times 2 \times 14 \times .03617 = 2.0255$ pounds, and the weight per square inch of area in cd is $\frac{2.0255}{4} = .5064$ pound.

566. According to Pascal's law, this weight (pressure) is transmitted equally in all directions, therefore, an extra weight of .5064 pound is imposed on every square inch of the bottom of bc; the area of this is $6 \times 6 = 36$ square inches, and the pressure on it is, therefore, $36 \times .5064 = 18.23$ pounds, due to water contained in cd; thus, we have a total pressure on bottom of vessel b of $13.02 + 18.23 = 31.25$ pounds, the same as in vessel a. As a result of the above law, there is also an upward pressure of .5064 pound acting on every square inch of the *top* of the enlarged part bc.

If an additional pressure of 10 pounds per square inch were applied to the upper surface of both vessels, the total pressure on each bottom would be $31.25 + (6 \times 6 \times 10) = 31.25 + 360 = 391.25$ pounds.

If this pressure were to be obtained by means of a weight placed on each piston (as shown in Figs. 103 and 104), we should have to put a weight of $6 \times 6 \times 10 = 360$ pounds on the piston in vessel a, and one of $2 \times 2 \times 10 = 40$ pounds on the piston in vessel b.

567. The General Law for the Downward Pressure upon the Bottom of any Vessel:

Rule 106.—*The pressure upon the bottom of a vessel containing a fluid is independent of the shape of the vessel, and is equal to the weight of a column of the fluid, the area of whose base is equal to that of the bottom of the vessel, and whose altitude is the distance between the bottom and the upper surface of the fluid plus the pressure per unit of area* *upon the upper surface of the fluid multiplied by the area of the bottom of the vessel.*

568. Suppose that the vessel b, in Fig. 105, were inverted, as shown in Fig. 106, the pressure upon the bottom will still be 0.868 pound per square inch, but it will require a weight of 3,490 pounds to be placed upon a piston at the upper surface to make the pressure on the bottom 391.25 pounds, instead of a weight of 40 pounds, as in the other case.

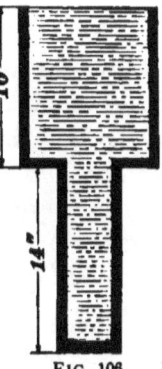

Fig. 106.

EXAMPLE.—A vessel filled with salt water, having a specific gravity of 1.03, has a circular bottom 13 inches in diameter. The top of the vessel is fitted with a piston 3 inches in diameter, on which is laid a weight of 75 pounds; what is the total pressure on the bottom, if the depth of the water is 18 inches?

SOLUTION.—Applying the rule, the weight of 1 cubic inch of the water is $\dfrac{62.5 \times 1.03}{1,728} = .037254$ lb.

$13 \times 13 \times .7854 \times 18 \times .037254 = 89.01$ pounds = the pressure due to the weight of the water.

$\dfrac{75}{3 \times 3 \times .7854} = 10.61$ pounds per square inch, due to the weight on the piston. $13 \times 13 \times .7854 \times 10.61 = 1,408.29$ pounds.

Total pressure $= 1,408.29 + 89.01 = 1,497.3$ pounds. Ans.

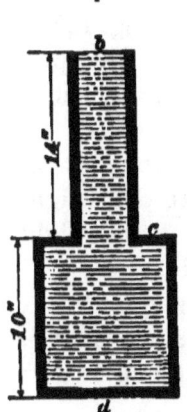

Fig. 107.

569. Upward Pressure.—In Fig. 107 is represented a vessel of exactly the same size as that shown in Fig. 106. There is no upward pressure on the surface c, due to the weight of the water in the large part $c\,d$, but there is an upward pressure on c, due to the weight of the water in the small part $b\,c$. The pressure per square inch, due to the weight of the water in $b\,c$, was found to be .5064 pound (see Art. **565**), the area of the upper surface c of the large part $c\,d$ is $(6 \times 6) - (2 \times 2) = 36 - 4 = 32$ sq. in., and the total upward pressure, due to the weight of the water, is $.5064 \times 32 = 16.2$ pounds.

MECHANICS.

If an additional pressure of 10 pounds per square inch were applied to a piston fitting in the top of the vessel, the total upward pressure on the surface c would be $16.2 + (32 \times 10) = 336.2$ pounds.

570. General Law for Upward Pressure:

Rule 107.—*The upward pressure on any submerged horizontal surface equals the weight of a column of the liquid whose base has an area equal to the area of the submerged surface, and whose altitude is the distance between the submerged surface and the upper surface of the liquid plus the pressure per unit of area on the upper surface of the fluid multiplied by the area of the submerged surface.*

EXAMPLE.—A horizontal surface 6 inches by 4 inches is submerged in a vessel of water 26 inches below the upper surface; if the pressure on the water is 16 pounds per square inch, what is the total upward pressure on the horizontal surface?

SOLUTION.—Applying the rule, we get $6 \times 4 \times 26 \times .03617 = 22.57$ pounds for the upward pressure, due to the weight of the water, and $6 \times 4 \times 16 = 384$ pounds for the upward pressure, due to the outside pressure of 16 pounds per square inch.

Therefore, the total upward pressure $= 384 + 22.57 = 406.57$ pounds. Ans.

571. Lateral (Sideways) Pressure.—Suppose the top of the vessel shown in Fig. 108 is 10 inches square, and that the projections at a and b are 1 inch × 1 inch and 10 inches long.

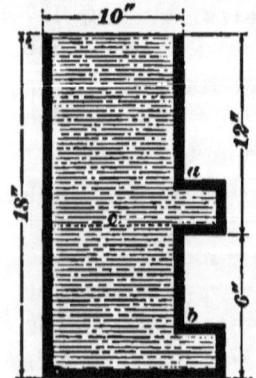

FIG. 108.

The pressure per square inch on the bottom of the vessel, due to the weight of the liquid, is $1 \times 1 \times 18 \times$ the weight of a cubic inch of the liquid.

The pressure at a depth equal to the distance of the upper surface b is $1 \times 1 \times 17 \times$ the weight of a cubic inch of the liquid.

Since both of these pressures are transmitted in every direction, they are also transmitted laterally (sideways), and the *pressure per unit of area on the projection b is a mean between the two*, and equals $1 \times 1 \times 17\frac{1}{2} \times$ the weight of a cubic inch of the liquid.

To find the lateral pressure on the projection a, imagine that the dotted line c is the bottom of the vessel, then the conditions would be the same as in the preceding case, except that the depth is not so great.

The lateral pressure per sq. in. on a is thus seen to be $1 \times 1 \times 11\frac{1}{2} \times$ the weight of a cubic inch of the liquid.

572. General Law for Lateral Pressure :

Rule 108.—*The pressure upon any vertical surface, due to the weight of the liquid, is equal to the weight of a column of the liquid whose base has the same area as the vertical surface, and whose altitude is the depth of the center of gravity of the vertical surface below the upper surface of the liquid.*

Any additional pressure is to be added as in the previous cases.

EXAMPLE.—A well 3 feet in diameter, and 20 feet deep, is filled with water; what is the pressure on a strip of the wall 1 inch wide, the top of which is 1 foot from the bottom of the well? What is the pressure on the bottom? What is the upward pressure per square inch, 2 feet 6 inches from the bottom?

SOLUTION.—Applying the rule, the area of the strip is equal to its length (= circumference of well) multiplied by its height. The length = $36 \times 3.1416 = 113.1'$; height = $1'$; hence, area of strip = $113.1 \times 1 = 113.1$ sq. in. Depth of center of gravity of strip = $(20 - 1)$ ft. + $\frac{1}{2}$ inch, the half width of strip, = $228\frac{1}{2}$ inches. Consequently, the total pressure on the strip = $113.1 \times 228.5 \times .03617 = 934.75$ lb.   Ans.

The pressure on each square inch of the strip would be $\frac{934.75}{113.1} = 8.265$ pounds, nearly.

$36 \times 36 \times .7854 \times 20 \times 12 \times .03617 = 8,836$ pounds = the pressure on the bottom.   Ans.

$20 - 2.5 = 17.5$. $1 \times 17.5 \times 12 \times .03617 = 7.596$ pounds = the upward pressure per square inch, 2 feet 6 inches from the bottom.   Ans.

573. The effects of lateral pressure are illustrated in Fig. 109. In the figure, c is a tall vessel having a stop-cock near its base, and arranged to float upon the water, as shown. When this vessel is filled with water, the lateral pressures at any two points of the surface of the vessel opposite to each other are equal. Being equal, and acting in opposite directions, they destroy each other, and no motion can result; but if the stop-cock be opened there will be no resistance to the pressure acting on that part, and the water will flow out; at the same time, the pressure on the corresponding part of the opposite side of the vessel remains, and this, being no longer balanced, causes the vessel to move backwards through the water in a direction opposite to that of the issuing jet.

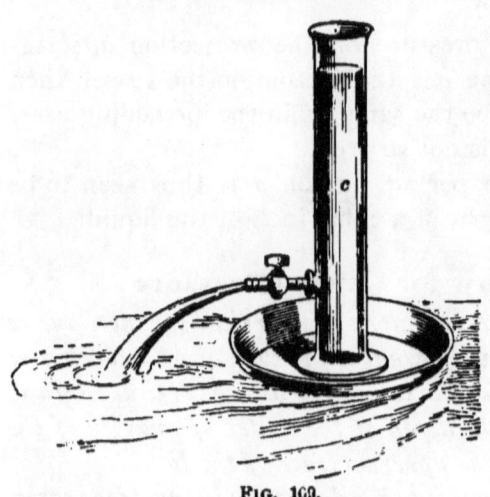

FIG. 109.

574. Since the pressure on the bottom of a vessel, due to the weight of the liquid, is dependent only upon the height of the liquid, and not upon the shape of the vessel, it follows

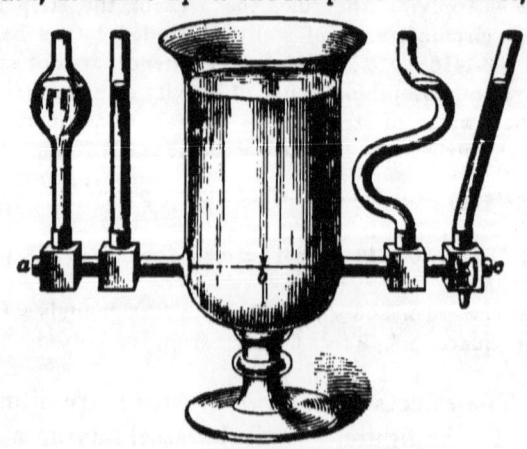

FIG. 110.

that if a vessel has a number of radiating tubes (see Fig. 110) the water in each tube will be on the same level, no matter what may be the shape of the tubes. For, if the water were higher in one tube than in the others, the downward pressure on the bottom, due to the height of the water in this tube, would be greater than that due to the height of the water in the other tubes. Consequently, the upward pressure would also be greater; the equilibrium would be destroyed, and the water would flow from this tube into the vessel, and rise in the other tubes until it was at the same level in all, when it would be in equilibrium. This principle is expressed in the familiar saying, *water seeks its level.*

575. The above principle explains why city water reservoirs are located on high elevations, and why water on leaving the hose nozzle spouts so high.

If there were no resistance by friction and air, the water would spout to a height equal to the level of the water in the reservoirs. If a long pipe were attached to the nozzle whose length was equal to the vertical distance between the nozzle and the level of the water in the reservoir, the water would just reach the end of the pipe. If the pipe were lowered slightly, the water would trickle out.

Fountains, canal locks, and artesian wells are examples of the application of this principle.

EXAMPLE.—The water level in a city reservoir is 150 feet from the level of the street; what is the pressure of the water per square inch on the hydrant?

SOLUTION.—Apply rule **106**, $1 \times 150 \times 12 \times .03617 = 65.106$ pounds per square inch.

NOTE.—In measuring the height of the water to find the pressure which it produces, the **vertical** height, or distance, between the level of the water and the point considered is always taken. This vertical height is called the **head.**

The weight of a column of water 1 in. square and 1 ft. high is $62.5 \div 144 = .434$ lb., very nearly. Hence, if the depth (head) be given, the pressure per square inch may be found by multiplying the depth in feet by .434. The constant .434 is the one ordinarily employed in practical calculations.

576. In Fig. 111, let the area of the piston *a* be 1 square inch, of *b*, 40 square inches. According to Pascal's law, 1 pound placed upon *a* will balance 40 pounds placed upon *b*.

Suppose that *a* moves downwards 10 inches, then, 10 cubic inches of water will be forced into the tube *b*. This will be

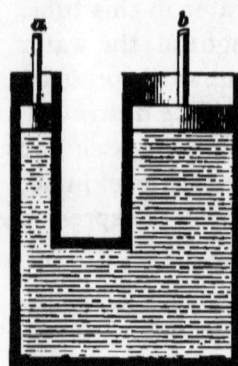

FIG. 111.

distributed over the entire area of the tube *b*, in the form of a cylinder whose cubical contents must be 10 cubic inches, whose base has an area of 40 square inches, and whose altitude must be $\frac{10}{40} = \frac{1}{4}$ of an inch; that is, a movement of 10 inches of the piston *a* will cause a movement of $\frac{1}{4}$ of an inch in the piston *b*.

Here is the old principle of machines: *The power multiplied by the distance through which it moves equals the weight multiplied by the distance through which it moves.*

Since, if 1 pound on the piston *a* represents the power *P*, the equivalent weight *W* on *b* may be obtained from the equation, $W \times \frac{1}{4} = P \times 10$, whence $W = 40\,P = 40$ lb.

577. Another familiar fact is also recognized, for the velocity ratio of *P* to *W* is $10 : \frac{1}{4}$, or 40; and, since in any machine the weight equals the power multiplied by the velocity ratio, $W = P \times 40$, and when $P = 1$, $W = 40$.

This principle is made use of in the hydraulic press represented in Fig. 112. As the man depresses the lever *O*, he forces down the piston *a* upon the water in the cylinder *A*. The water is forced through the bent tube *d* into the cylinder in which the large ram, or plunger, *C* works, and causes it to rise, thus lifting the platform *K*, and compressing the bales. If the area of *a* be 1 sq. in., and that of *c* be 100 sq. in., the velocity ratio will be 100.

If the length of the lever between the hand and the fulcrum is 10 times the length between the fulcrum and the piston *a*, the velocity ratio of the lever will be 10, and the total velocity ratio of the hand to the piston *C* will be $10 \times 100 = 1,000$.

Hence, a force of 100 pounds applied by the hand will raise $100 \times 1{,}000 = 100{,}000$ pounds. But, if the average move-

FIG. 112.

ment of the hand per stroke is 10 inches, it will require $\frac{1{,}000}{10} = 100$ strokes to raise the platform 1 inch, and it is again seen that what is gained in power is lost in speed.

578. Applications of this principle are seen in the hydraulic machines used for forcing locomotive drivers on their axles, etc., and for testing the strength of boiler shells.

EXAMPLE.—A suspended vertical cylinder is tested for the tightness of its heads by filling it with water. A pipe whose inside diameter is ¼ of an inch, and whose length is 20 feet, is screwed into a hole in the upper head, and then filled with water. What is the pressure per square inch on each head, if the cylinder is 40 inches in diameter and 60 inches long?

SOLUTION.—Area of heads $= 40^2 \times .7854 = 1{,}256.64$ sq. in.

Pressure per square inch on the bottom head, due to the weight of the water in the cylinder, $= 1 \times 60 \times .03617 = 2.17$ pounds.

$(\tfrac{1}{4})^2 \times .7854 = .04909$ sq. in., the area of the pipe.

$.04909 \times 20 \times 12 \times .03617 = .426$ pound = the weight of water in the pipe = the pressure on a surface area of $.04909$ sq. in.

The pressure per square inch, due to the water in the pipe, is $\frac{1}{.04909} \times .426 = 8.68$ pounds per square inch upon the upper head. Ans.

The total pressure per square inch on the lower head is $8.68 + 2.17 = 10.85$ pounds. Ans.

EXAMPLE.—In the last example, if the pipe be fitted with a piston weighing $\tfrac{1}{4}$ of a pound, and a 5-pound weight be laid upon it, what is the pressure per sq. in. upon the upper head?

SOLUTION.—In addition to the pressure of $.426$ pound on the area of $.04909$ sq. in., there is now an additional pressure upon this area of $5 + \tfrac{1}{4} = 5.25$ pounds, and the total pressure upon this area is $.426 + 5.25 = 5.676$ pounds. The pressure per square inch is $\frac{1}{.04909} \times 5.676 = 115.6$ pounds. Ans.

EXAMPLE.—In Fig. 113, the plunger A, is 10" in diameter, and is forced

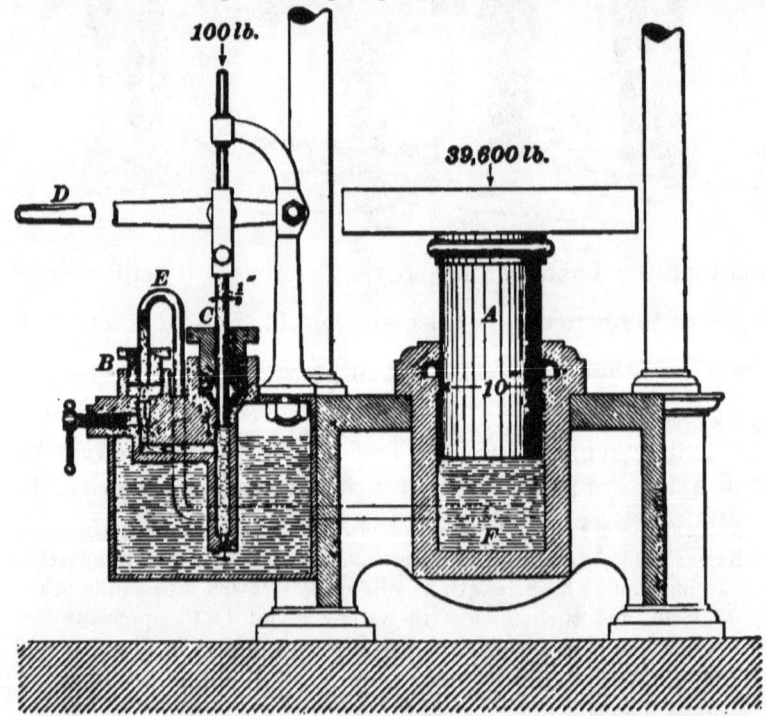

FIG. 113.

outwards by means of a small pump B, which supplies the press cylinder with water. The plunger C of the pump B is $\frac{1}{2}$" in diameter. If a force of 100 lb. be applied to C by means of the lever D, how great a weight can the plunger A raise, if the plunger itself weighs 400 lb. ?

SOLUTION.—First find the pressure per sq. in. which the plunger C exerts upon the water. Area of $C = (\frac{1}{2})^2 \times .7854 = .19635$ sq. in. Since C exerts a pressure of 100 lb. upon an area of .19635 sq. in., it will exert a pressure of $100 \times \frac{1}{.19635}$ lb. upon an area of 1 sq. in. This pressure is transmitted by the water in the tube E to the cylinder F, where it forces the plunger A upwards as the water is forced into F. The pressure per sq. in. on the bottom of A is the same as that exerted by C. Hence, the total pressure is $10^2 \times .7854 \times \frac{100}{.19635} = 40,000$ lb. This result, less the weight of A, equals the load lifted. Therefore, $40,000 - 400 = 39,600$ lb. Ans.

579. In these examples on the hydraulic press, no allowance has been made for the power lost in overcoming the friction between the cup leathers and the plunger; this varies according to the condition of the leathers, and, of course, the smoothness of the plunger; when the leathers are in good condition the loss is about 5% of the total pressure on the ram; when the leathers are old, stiff, and dirty the loss may amount to 15% or more.

BUOYANT EFFECTS OF WATER.

580. In Fig. 114 is shown a 6-inch cube, entirely submerged in water. The lateral pressures are equal and in opposite directions. The upward pressure acting on the lower surface of the cube $= 6 \times 6 \times 21 \times .03617$; the downward pressure acting on the top of the cube $= 6 \times 6 \times 15 \times .03617$, and the difference $= 6 \times 6 \times 6 \times .03617 =$ the volume of the cube in cubic inches $\times$ the weight of one cubic inch of water. That is, the upward pressure exceeds the downward pressure by the weight of a volume of water equal to the volume of the body.

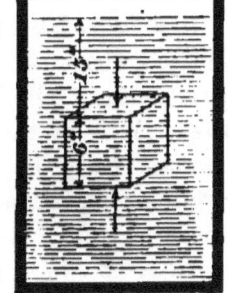

FIG. 114.

This excess of upward pressure over the downward pressure

acts against gravity; that is, the water presses the body *up* with a greater force than it presses it *down;* consequently, *if a body is immersed in a fluid, it will lose in weight an amount equal to the weight of the fluid it displaces.* This is called the **principle of Archimedes,** because it was first stated by him.

581. This principle may be experimentally demonstrated with the beam scales, as shown in Fig. 115.

From one scale pan suspend a hollow cylinder of metal *t*, and below that a solid cylinder *a* of the same size as the hollow part of the upper cylinder. Put weights in the other scale pan until they exactly balance the two cylinders. If *a* be immersed in water, the scale pan containing the weights will descend, showing that *a* has lost some of its weight. Now fill *t* with water, and the volume of water that can be poured into *t* will equal that displaced by *a*. The scale pan that contains the weights will gradually rise until *t* is filled, when the scales balance again.

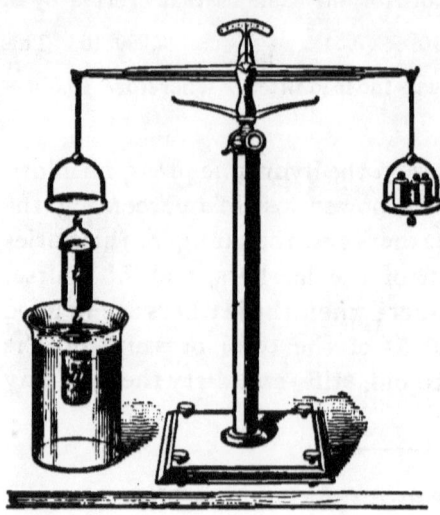

FIG. 115.

If a body be lighter than the liquid in which it is immersed, the upward pressure will cause it to rise, and project partly out of the liquid, until the weight of the body and the weight of the liquid displaced are equal. If the immersed body be heavier than the liquid, the downward pressure plus the weight of the body will be greater than the upward pressure, and the body will fall downwards until it touches bottom or meets an obstruction. If the weights of equal volumes of the liquid and the body are

MECHANICS.

equal, the body will remain stationary, and be in equilibrium in any position or depth beneath the surface of the liquid.

582. An interesting experiment in confirmation of the above facts may be performed as follows: Drop an egg into a glass jar filled with fresh water. The mean density of the egg being a little greater than that of water, it will fall to the bottom of the jar. Now, dissolve salt in the water, stirring it so as to mix the fresh and salt water. The salt water will presently become denser than the egg, and the egg will rise. Now, if fresh water be poured in until the egg and water have the same density, the egg will remain stationary in any position that it may be placed below the surface of the water.

EXAMPLES FOR PRACTICE.

1. Suppose a cylinder to be filled with water and placed in an upright position. If the diameter of the cylinder be 19", and its total length inside be 26", what will be the total pressure on the bottom when a pipe $\frac{1}{2}$" in diameter and 12 ft. long is screwed into the cylinder head and filled with water? The pipe is vertical. Ans. 1,743.2 lb.

2. In the last example, what is the total pressure against the upper head? Ans. 1,476.6 lb.

3. In example 1, a piston is fitted to the upper end of the pipe, and an additional force of 10 lb. is applied to the water in the pipe. (*a*) What is the total pressure on the bottom of the cylinder? (*b*) On the upper head?   Ans. { (*a*) 16,184 lb. (*b*) 15,906 lb.

4. In example 3, what is the pressure per square inch in the pipe 2" from the upper cylinder head? Ans. 56.0656 lb. per sq. in.

5. A water tower 80 feet high is filled with water. A pipe 4" in diameter is so connected to the side of the tower that its center is 3 feet from the bottom. If the pipe is closed by a flat cover, what is the total pressure against the cover? Ans. 420 lb.

6. In the last example, what is the upward pressure per square inch 10 feet from the bottom of the tower?
Ans. 30.3828 lb. per sq. in.

7. A cube of wood, one edge of which measures 3 feet, is sunk until the upper surface is 40 feet below the level of the water; what is the total force which tends to move the cube upwards? Ans. 1,687.5 lb.

MECHANICS.

PNEUMATICS.

583. Pneumatics is that branch of mechanics which treats of the properties of gases.

584. The most striking feature of all gases is their extreme expansibility. *If we inject a portion of gas, however small it may be, into a vessel, it will expand and fill that vessel.* If a bladder or football be partly filled with air, and placed under a glass jar (called a receiver), from which the air has been exhausted, the bladder or football will immediately expand, as shown in Fig. 116. The force which a gas always exerts when confined in a limited space is called **tension**. The word *tension* in this case means pressure, and is only used in this sense in reference to gases.

FIG. 116.

585. As *water* is the most common type of fluids so *air* is the most common type of gases. It was supposed by the ancients that air had no weight, and it was not until about the year 1650 that it was proven that air really had weight. A cubic inch of air, under ordinary conditions, weighs .31 grain, nearly. At a temperature of 32° and a pressure of 14.7 pounds per square inch, the ratio of the weight of air to water is about 1 : 774; that is, air is only $\frac{1}{774}$ as heavy as water. In Art. **580,** it was shown that if a body were immersed in water, and weighed less than the volume of water displaced, the body would rise and project partly out of the water. The same is true to a certain extent of air. If a vessel made of light material be filled with a gas lighter than air, so that the total weight of the vessel and

MECHANICS.   223

gas is less than the air they displace, the vessel will rise. It is on this principle that balloons are made.

586. Since air has weight, it is evident that the enormous quantity of air that constitutes the atmosphere must exert a considerable pressure upon the earth. This is easily proven by taking a long glass tube closed at one end, and filling it with mercury. If the finger be placed over the open end so as to keep the mercury from running out, and the tube inverted and placed in a cup of mercury, as shown in Fig. 117, the mercury will fall, then rise, and after a few oscillations will come to rest at a height above the top of the mercury in the glass equal to about 30 inches. This height will always be the same under the same atmospheric conditions. Now, if the atmosphere has weight, it must press upon the upper surface of the mercury in the glass with equal intensity upon every square unit, except upon that part of the surface occupied by the tube. In order that there may be equilibrium, the weight of the mercury in the tube must be equal to the pressure of the air upon a portion of the surface of the mercury in the glass, equal in area to the inside of the tube. Suppose that the area of the inside of the tube is 1 square inch, then, since mercury is 13.6 times as heavy as water, the weight of the mercury column is $.03617 \times 13.6 \times 30 = 14.7574$ lb. The actual height of the mercury is a little less than 30 inches, and the actual weight of a cubic inch of distilled water is a little less than .03617 lb. When these considerations are taken into account, the average weight of the mercurial

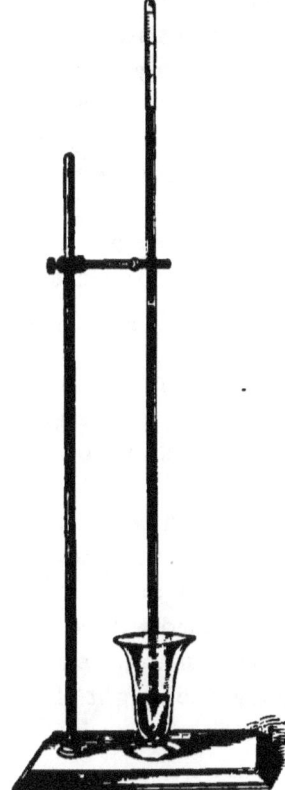

FIG. 117.

column at the level of the sea, when the temperature is 60°, is 14.69 lb., or practically 14.7 lb. Since this weight, when exerted upon 1 square inch of the liquid in the glass, just produces equilibrium, it is plain that the pressure of the outside air is 14.7 lb. upon every square inch of surface.

587. Vacuum.—The space between the upper end of the tube and the upper surface of the mercury is called a **vacuum**, meaning that it is an entirely empty space, and does not contain any substance, solid, liquid, or gaseous. If there were a gas of some kind there, no matter how small the quantity might be, it would expand, filling the space, and its tension would cause the column of mercury to fall and become shorter, according to the amount of gas or air present. The space is then called a **partial vacuum**. If the mercury fell 1 inch, so that the column was only 29 inches high, we should say, in ordinary language, that there were *29 inches of vacuum*. If it fell 8 inches, we should say that there were 22 inches of vacuum; if it fell 16 inches, we should say that there were 14 inches of vacuum, and so on. Hence, when the vacuum gauge of a condensing engine shows 26 inches of vacuum, there is enough air in the condenser to produce a pressure of $\frac{30-26}{30} \times 14.7 = \frac{4}{30} \times 14.7 = 1.96$ lb. per sq. in. In all cases where the mercury column is used to measure a vacuum, the height of the column in inches gives the number of inches of vacuum. Were the column only 5" high, the vacuum would be 5".

If the tube had been filled with water instead of mercury, the height of the column of water to balance the pressure of the atmosphere would have been $30 \times 13.6 = 408$ inches $= 34$ feet. This means that, if a tube be filled with water, inverted, and placed in a dish of water in a manner similar to the experiment made with the mercury, the height of the column of water will be 34 feet.

588. The **barometer** is an instrument used for measuring the pressure of the atmosphere. There are two kinds

MECHANICS.   225

in general use—the mercurial barometer and the aneroid barometer. The **mercurial barometer** is shown in Fig. 118. The principle is the same as the inverted tube shown in Fig. 117. In this case the tube and the cup at the bottom are protected by a brass or iron casing. At the top of the tube is a graduated scale which can be read to $\frac{1}{1000}$ of an inch by means of a vernier. Attached to the casing is an accurate thermometer for determining the temperature of the outside air at the time the barometric observation is taken. This is necessary, since mercury expands when the temperature is increased, and contracts when the temperature falls; for this reason a standard temperature is assumed, and all barometer readings are reduced to this temperature. This standard temperature is usually taken at 32° F., at which temperature the height of the mercurial column is 30 inches. Another correction is made for the altitude of the place above sea-level, and a third correction for the effects of capillary attraction.

589. In Fig. 119 is an illustration of an **aneroid barometer.** These instruments are made in various sizes, from the size of a watch up to 8 or 10 inches in diameter. They consist of a cylindrical box of metal with a top of thin, elastic corrugated metal. The air is exhausted from the box. When the atmospheric pressure increases, the top is pressed inwards, and when it is diminished, the top is pressed outwards by its own elasticity, aided by a spring beneath. These movements of the cover are transmitted and multiplied by a combination of delicate levers, which act upon an index hand, and cause it to move either to the right or left, over a graduated scale. These barometers are self-correcting (compensated) for variations in temperature. They are very portable, occupying

FIG. 118.

but a small space, and are so delicate that they are said to show a difference in the atmospheric pressure when trans-

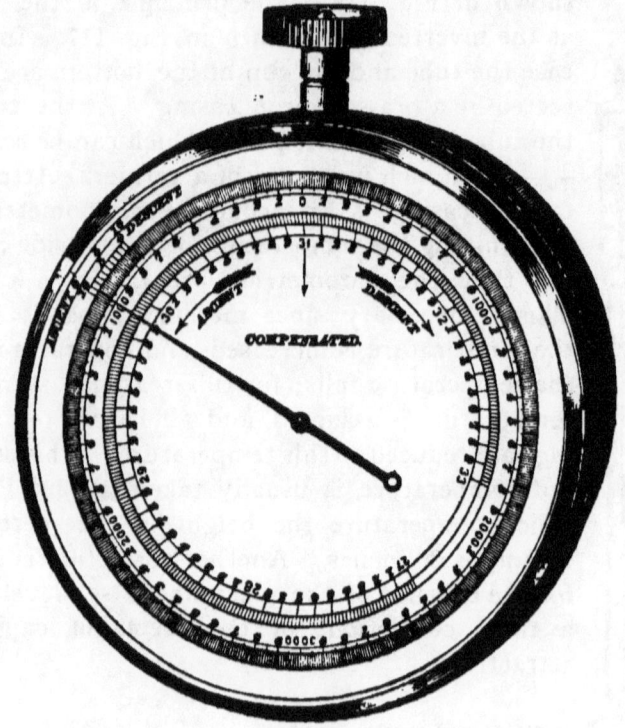

FIG. 119.

ferred from the table to the floor. The mercurial barometer is the standard.

590. With air, as with water, the lower we get the greater the pressure, and the higher we get the less the pressure. At the level of the sea, the height of the mercurial column is about 30 inches; at 5,000 feet above the sea, it is 24.7 inches; at 10,000 feet above the sea, it is 20.5 inches; at 15,000 feet, it is 16.9 inches; at 3 miles, it is 16.4 inches, and at 6 miles above the sea-level, it is 8.9 inches.

The density or weight of the atmosphere also varies with the altitude; that is, a cubic foot of air at an elevation of 5,000 feet above the sea-level will not weigh as much as a cubic foot at sea-level. This is proven conclusively by the fact that at a height of 3½ miles the mercurial column measures but 15 inches, indicating that half the weight of the entire atmosphere is below that. It is known that the height of the earth's atmosphere is at least 50 miles; hence, the air just before reaching the limit must be in an exceedingly rarefied state. It is by means of barometers that great heights are measured. The aneroid barometer has the heights marked on the dial, so that they can be read directly. With the mercurial barometer, the heights must be calculated from the reading.

591. The atmospheric pressure is everywhere present, and presses all objects in all directions with equal force. If a book is laid upon the table, the air presses upon it in every direction with an equal average force of 14.7 pounds per square inch. It would seem as though it would take considerable force to raise a book from the table, since, if the size of the book were 8 inches by 5 inches, the pressure upon it would be $8 \times 5 \times 14.7 = 588$ lb.; but there is an equal pressure *beneath* the book which counteracts the pressure on the top. It would now seem as though it would require a great force to open the book, since there are two pressures of 588 pounds each acting in opposite directions and tending to crush the book; so it would, but for the fact that there is a layer of air between each leaf acting upwards and downwards with a pressure of 14.7 pounds per square inch. If two metal plates be made as perfectly smooth and flat as it is possible to get them, and the edge of one be laid upon the edge of the other, so that one may be slid upon the other and thus exclude the air, it will take an immense force, compared to the weights of the plates, to separate them. This is because the full pressure of 14.7 pounds per square inch is then exerted upon each plate, with no counteracting equal pressure between them.

If a piece of flat glass be laid upon a flat surface that has been previously moistened with water, it will require considerable force to separate them; this is because the water helps to fill up the pores in the flat surface and glass, and thus creates a partial vacuum between the glass and the surface, thereby reducing the counter pressure beneath the glass.

592. Tension of Gases.—In Fig. 117, the space above the column of mercury was said to be a vacuum, and it was also said that if any gas or air were present, it would expand, its tension forcing the column of mercury downwards. If sufficient gas be admitted to cause the mercury to stand at 15 inches, the tension of the gas is evidently $\frac{14.7}{2} = 7.35$ lb. per square inch, since the pressure of the outside air, 14.7 lb. per square inch, now balances only 15 instead of 30 inches of mercury; that is, it balances only half as much as it would if there were no gas in the tube; therefore, the pressure (tension) of the gas in the tube is 7.35 pounds. If more gas be forced into the tube until the top of the mercurial column is just level with the mercury in the cup, the gas in the tube will then have a tension equal to the outside pressure of the atmosphere. Suppose that the bottom of the tube is fitted with a piston, and that the total length of the inside of the tube is 36 inches. If the piston be shoved upwards so that the space occupied by the gas is 18 inches long instead of 36 inches, the temperature remaining the same as before, it will be found that the tension of the gas within the tube is 29.4 lb. It will be noticed that the volume occupied by the gas is only half that in the tube before the piston was moved, while the pressure is twice as great, since $14.7 \times 2 = 29.4$ lb. If the piston be shoved up, so that the space occupied by the gas is only 9 inches, instead of 18 inches, the temperature still remaining the same, the pressure will be found to be 58.8 pounds per square inch. The volume has again been reduced one-half, and the pressure increased two-fold, since $29.4 \times 2 = 58.8$ lb. The space now occupied by the gas is 9 inches long, whereas, before the piston was moved, it was 36 inches long; as the tube was assumed to be of uniform diameter throughout its length, the volume is now $\frac{9}{36} = \frac{1}{4}$ of its original volume, and its pressure is $\frac{58.8}{14.7} = 4$ times its original pressure. Moreover, if the temperature of the confined gas remains the same, the pressure and volume will always vary in a similar way.

593. The law which states these effects is called *Mariotte's law*, and is as follows:

Mariotte's Law.—*The temperature remaining the same, the volume of a given quantity of gas varies inversely as the pressure.*

The meaning of the law is this: If the volume of a gas be diminished to $\frac{1}{2}$, $\frac{1}{3}$, $\frac{1}{5}$, etc., of its former volume, the tension will be increased 2, 3, 5, etc., times, or if the outside pressure be increased 2, 3, 5, etc., times, the volume of the gas will be diminished to $\frac{1}{2}$, $\frac{1}{3}$, $\frac{1}{5}$, etc., of its original volume, the temperature remaining constant. It also means that if a gas is under a certain pressure, and this pressure is diminished to $\frac{1}{2}$, $\frac{1}{4}$, $\frac{1}{10}$, etc., of its original intensity, the volume of the confined gas will be increased 2, 4, 10, etc., times—its tension decreasing at the same rate.

Suppose 3 cubic feet of air to be under a pressure of 60 pounds per square inch in a cylinder fitted with a movable piston, then the product of the volume and pressure is $3 \times 60 = 180$. Let the volume be increased to 6 cubic feet, then the pressure will be 30 pounds per square inch, and $30 \times 6 = 180$, as before. Let the volume be increased to 24 cubic feet; it is then $\frac{24}{3} = 8$ times its original volume, and the pressure is $\frac{1}{8}$ of its original pressure, or $60 \times \frac{1}{8} = 7\frac{1}{2}$ lb., and $24 \times 7\frac{1}{2} = 180$, as in the two preceding cases. It will now be noticed that if a gas be enclosed within a confined space, and allowed to expand without losing any heat, *the product of the pressure and the corresponding volume for any one position of the piston is the same as for any other position.* If the piston were forced inwards so as to compress the air, the same results would be obtained.

594. If the volume of the vessel and the pressure of the gas are known, and it is desired to know the pressure after the first volume has been changed:

Rule 109.—*Divide the product of the first, or original, volume and pressure by the new volume; the result will be the new pressure.*

Or, let p = original pressure;
p_1 = final pressure;
v = volume corresponding to the pressure p;
v_1 = volume corresponding to the pressure p_1.

Then, $$p_1 = \frac{pv}{v_1}.$$

EXAMPLE.—At the point of cut-off in a steam engine, the amount of steam in the cylinder is 862 cu. in. The pressure at this point is 120 lb. per sq. in. What will be the pressure of the steam when the piston has reached the end of its stroke, and the volume is 1,800 cu. in.?

SOLUTION.—Applying the rule, $\frac{862 \times 120}{1,800} = 57.47$ lb. per sq. in. Ans.

595. If it is required to determine the volume after a change in the pressure:

Rule 110.—*Divide the product of the original volume and pressure by the new pressure; the result will be the new volume.*

Or, using the same letters as before,

$$v_1 = \frac{pv}{p_1}.$$

EXAMPLE.—At the commencement of compression, the volume of the steam is 380 cu. in., and the pressure is 18 lb. per sq. in. At the end of compression, the pressure is 112 lb. per sq. in. What is the final volume?

SOLUTION.—Applying the rule, $\frac{380 \times 18}{112} = 61.07$ cu. in. Ans.

EXAMPLE.—A vessel contains 10 cu. ft. of air at a pressure of 15 lb. per sq. in., and has 25 cu. ft. of air of the same pressure forced into it; what is the resulting pressure?

SOLUTION.—The original volume = 10 + 25 = 35 cu. ft. The original pressure is 15 lb. per sq. in. The final volume is 10 cu. ft. Hence, applying rule **109**, $\frac{35 \times 15}{10} = 52.5$ lb. per sq. in. Ans.

MECHANICS.

It must be remembered that in the preceding examples the temperature is supposed to remain constant.

EXAMPLES FOR PRACTICE.

1. A vessel contains 25 cubic feet of gas, at a pressure of 18 lb. per sq. in.; if 125 cu. ft. of gas having the same pressure are forced into the vessel, what will be the resulting pressure? Ans. 108 lb. per sq. in.

2. The volume of steam in the cylinder of a steam engine at cut-off is 1.35 cu. ft., and the pressure is 85 lb. per sq. in. The pressure at the end of the stroke is 25 lb. per sq. in. What is the new volume?
Ans. 4.59 cu. ft.

3. A receiver contains 180 cu. ft. of gas, at a pressure of 20 lb. per sq. in.; if a vessel holding 12 cu. ft. be filled from the larger vessel until its pressure is 20 lb. per sq. in., what will be the pressure in the larger vessel? Ans. $18\frac{3}{4}$ lb. per sq. in.

4. A spherical shell has a part of the air within it removed, forming a partial vacuum; if the outside diameter of the shell is 18", and the pressure of the air within is 5 lb. per sq. in., what is the total pressure tending to crush the shell? Ans. 9,873.42 lb.

PNEUMATIC MACHINES.

596. The Air Pump.—*The air pump is an instrument for removing air from a given space.* A section of the

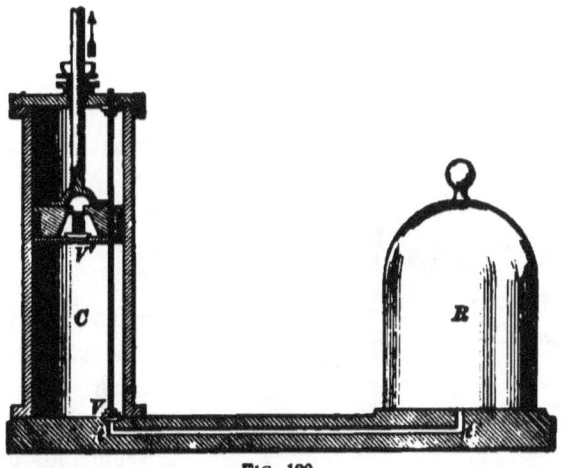

FIG. 120

principal parts is shown in Fig. 120, and the complete instrument in Fig. 121. The closed vessel R is called the **receiver,** and the space which it encloses is that from which it is desired to remove the air. It is usually made of glass, and the edges are ground so as to be perfectly air-tight. When made in the form shown, it is called a **bell-jar receiver.** The receiver rests upon a horizontal plate, in the center of which is an opening communicating with the pump cylinder C by means of the passage $t\ t$. The pump piston fits the cylinder accurately, and has a valve V' opening upwards. Where the passage $t\ t$ joins the cylinder, is another valve V, also opening upwards. When the piston is raised, the valve V' closes, and, since no air can get into the cylinder from above, the piston leaves a vacuum behind it. The pressure upon V being now removed, the tension of the air in the receiver R causes V to rise; the air in the receiver and passage $t\ t$ then expands so as to occupy the additional space provided by the upward movement of the piston.

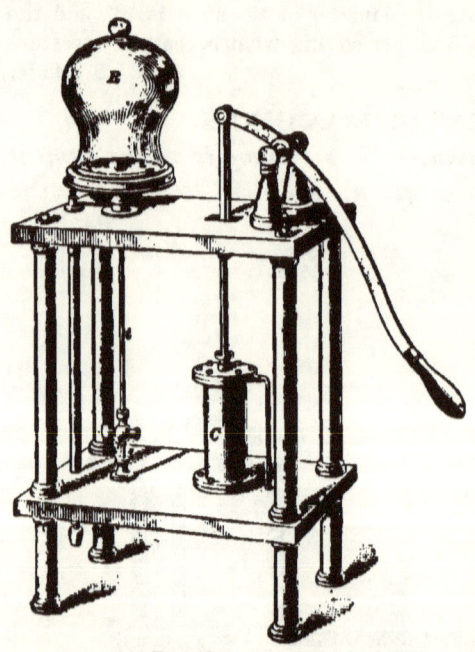

Fig. 121.

The piston is now pushed down, the valve V closes, the valve V' opens, and the air in C escapes. The lower valve V is sometimes supported, as shown in Fig. 120, by a metal rod passing through the piston, and fitting it somewhat tightly. When the piston is raised or lowered, this rod moves with it. A button near the upper end of the rod confines its motion within very narrow limits, the piston sliding upon the rod during the greater part of the journey.

In the complete form of the instrument shown in Fig. 121, communication between receiver and pump is made by means of the tube t.

597. Degrees and Limits of Exhaustion.—Suppose that the volume of R and t together is four times that of C, Fig. 121, and that there are say 200 grains of air in R and t, and 50 grains in C when the piston is at the top of the cylinder. At the end of the first stroke, when the piston is again at the top, 50 grains of air in the cylinder C will have been removed, and the 200 grains in R and t will occupy the space R, t, and C. The ratio between the sum of the spaces R and t, and the total space $R + t + C$ is $\frac{4}{5}$; hence, $200 \times \frac{4}{5} = 160$ grains = the weight of air in R and t after the first stroke. After the second stroke, the weight of the air in R and t would be $(200 \times \frac{4}{5}) \times \frac{4}{5} = 200 \times (\frac{4}{5})^2 = 200 \times \frac{16}{25} = 128$ grains. At the end of the third stroke the weight would be $[200 \times (\frac{4}{5})^2] \times \frac{4}{5} = 200 \times (\frac{4}{5})^3 = 200 \times \frac{64}{125} = 102.4$ grains. At the end of n strokes the weight would be $200 \times (\frac{4}{5})^n$. It is evident that *it is impossible to remove all of the air that is contained in R and t by this method.* It requires an exceedingly good air pump to reduce the tension of the air in R to $\frac{1}{60}$ of an inch of mercury. When the air has become so rarefied as this, the valve V' will not lift, and, consequently, no more air can be exhausted.

598. Magdeburg Hemispheres. —By means of the two hemispheres shown in Fig. 122, it can be proven that the atmosphere presses upon a body equally in all directions. They were invented by Otto Von Guericke, of Magdeburg, and are called the *Magdeburg hemispheres.* One of the hemispheres is provided with a stop-cock, by which it can be screwed on to an air pump. The edges fit accurately, and are well greased, so as to be air-tight. As long as the hemispheres contain air, they can be separated with no difficulty; but when the air in the interior is pumped out by means of an

Fig. 122.

MECHANICS.

air pump, they can be separated only with great difficulty. The force required to separate them will be equal to the area of the largest circle of the hemisphere in square inches multiplied by 14.7 pounds.

This force will be the same in whatever position the hemisphere may be held, thus proving that the pressure of air upon it is the same in all directions.

NOTE.—A theoretically perfect vacuum is sometimes called a *Torricellian vacuum*.

599. The Weight Lifter.—The pressure of the atmosphere is very clearly shown by means of an apparatus like that illustrated in Fig. 123. Here a cylinder fitted with a piston is held in suspension by a chain. At the top of the cylinder is a plug A which can be taken out. This plug is removed, the piston pushed up (the force necessary being equal to the weight of the piston and rod B) until it touches the cylinder head. The plug is then screwed in, and the piston will remain at the top until a weight has been hung on the rod equal to the number of pounds obtained by multiplying the number of square inches in the piston by 14.7 pounds, and subtracting therefrom the weight of the piston and rod in pounds.

FIG. 123.

If a force sufficiently great were employed to pull the piston downwards, and then any less weight were attached, as shown in Fig. 123, the piston would ascend and carry the weight up with it.

600. Suppose the weight to be removed, and the piston to be supported, say midway of the length of the cylinder. Let the plug be removed, and air admitted above the piston, then screw the plug back into its place; if the piston be shoved upwards, the further up it goes the greater will be the force necessary to push it, on account of the compression of the air. If the piston is of large diameter it will also require a great force to pull it out of the cylinder, as a little consideration will show. For example, let the diameter of the piston be 20 inches, the length of the cylinder 36 inches, and the weight of the piston and rod 100 pounds. If the piston is in the middle of the cylinder, there will be 18 inches of space above it, and 18 inches of space below it. The area of the piston is $20^2 \times .7854 = 314.16$ square inches, and the atmospheric pressure upon it is $314.16 \times 14.7 = 4,618$ lb., nearly. In order to shove the piston upwards 9 inches, the pressure upon it must be twice as great, or 9,236 pounds, and to this must be added the weight of the piston and rod, or $9,236 + 100 = 9,336$ lb. The force necessary to cause the piston to move upwards 9 inches would then be $9,336 - 4,618 = 4,718$ lb. Now, suppose the piston to be moved downwards until it is just on the point of being pulled out of the cylinder. The volume above it will then be twice as great as before, and the pressure one-half as great, or $4,618 \div 2 = 2,309$ lb. The total upward pressure will be the pressure of the atmosphere less the weight of the piston and rod, or $4,618 - 100 = 4,518$ lb., and the force necessary to pull it downwards to this point will be $4,518 - 2,309 = 2,209$ lb.

·601. Air Compressors.—For many purposes compressed air is preferable to steam or other gas for use as a motive power. In such cases **air compressors** are used to compress the air. These are made in many forms, but the most common one is to place a cylinder, called the **air cylinder,** in front of the cross-head of a steam engine, so that the piston of the air cylinder can be driven by attaching its piston rod to the cross-head, in a manner similar to a steam pump. A cross-section of the air cylinder of a compressor of this kind is shown in Fig. 124, in which A is the piston, and B is the piston rod, driven by the cross-head of a steam engine, not shown in the figure. Both ends of the lower half of the cylinder are fitted with inlet valves D

and D' which allow the air to enter the cylinder, and both ends of the upper half are fitted with discharge valves F and

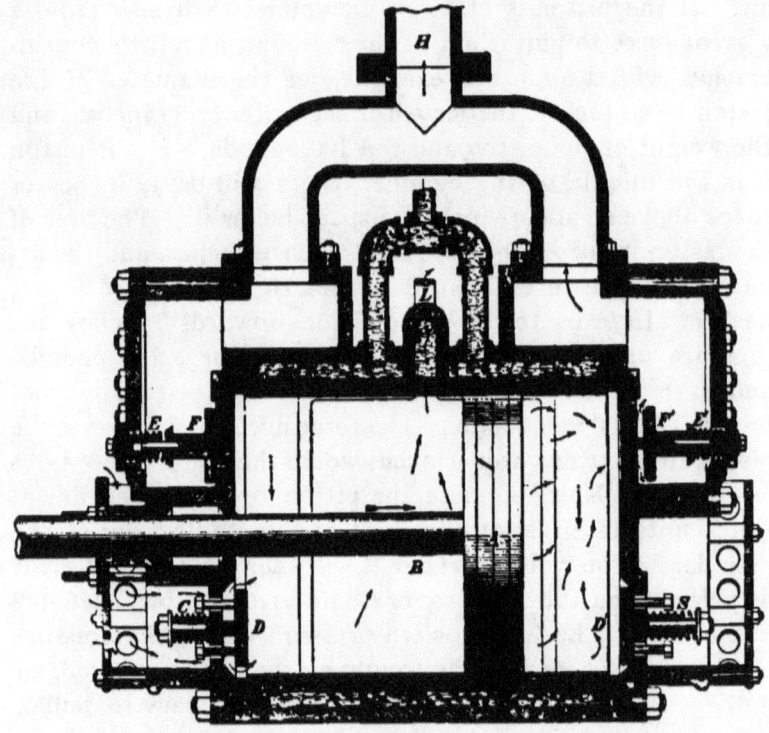

Fig. 124.

F' which allow the air to escape from the cylinder after it has been compressed to the required pressure.

602. Suppose the piston A to be moving in the direction of the arrow, then all the inlet valves D in the lower half of the left-hand end of the cylinder from which the piston is moving will be forced inwards by the pressure of the atmosphere, which overcomes the resistance of the spring C tending to keep the valve on its seat, thus allowing the air to rush into the cylinder. On the other side of the piston the air is being compressed, and, consequently, it forces the inlet valves D' in the lower half of the right-hand end of the cylinder to their seats against the resistance of their springs S. In the upper half of the right-hand end of the cylinder the discharge valves F' are opened by the compressed air against the resistance of the springs E', and in the upper half of the left-hand end of the cylinder the discharge valves F are pressed against their seats by the springs E, the tension of these springs being so adjusted that the valves can not be forced from their seats until the air on the other side has been compressed to the required pressure per square inch. For example, suppose it is desired to compress the air to 59 pounds per square inch, and we wish to find at what point of the stroke the discharge valves will open, they having been set for this pressure. Now, 59 pounds per square inch = 4 atmospheres very nearly; hence, the volume must be $\frac{1}{4}$ of the volume at the beginning of the stroke, or the valves will open when the piston has traveled $\frac{3}{4}$ of its stroke.

The air, after being discharged from the cylinder, passes out through the discharge pipe H, and from thence is conveyed to its destination.

603. Hero's Fountain.—Hero's fountain derives its name from its inventor, Hero, who lived at Alexandria 120 B. C.; it is shown in Fig. 125. It depends for its operation upon the elastic properties of air. It consists of a brass dish A, and two glass globes B and C. The dish communicates with the lower part of the globe C by means of a long tube D, and another tube E connects the two globes. A third tube passes through the dish A to the lower part of the globe B. This last tube being taken out the globe B is partially filled with water, the tube is then replaced, and water is poured into the dish. The

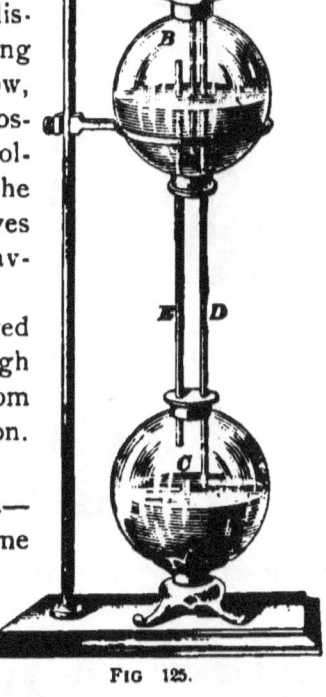

FIG. 125.

water flows through the tube D into the lower globe, and expels the air, which is forced into the upper globe. The air thus compressed acts upon the water and makes it jet out as represented in the figure. Were it not for the resistance of the atmosphere and friction, the water would rise to a height above the water in the dish equal to the difference of the level of the water in the two globes.

604. The Siphon.—The action of the *siphon* illustrates the effect of atmospheric pressure. It is simply a bent tube of unequal branches, open at both ends, and is used to convey a liquid from a higher point to a lower, over an intermediate point higher than either. In Fig. 126, A and B are two vessels, B being lower than A, and $A C B$ is the bent tube, or siphon. Suppose this tube to be filled with water and placed in the vessels as shown, with the short branch $A C$ in the vessel A. The water will flow from the vessel A into B, as long as the level of the water in B is below the level of the water in A, and the level of the water in A is above the lower end of the tube $A C$. The atmospheric pressure upon the surfaces of A and B tends to force the water up the tubes $A C$ and $B C$. When the siphon is filled with water each of these pressures is counteracted in part by the pressure of the water in that branch of the siphon which is immersed in the water upon which the pressure is exerted. The atmospheric pressure opposed to the weight of the longer column of water will, therefore, be more resisted than that opposed to the weight of the shorter column; consequently, the pressure exerted upon the shorter column

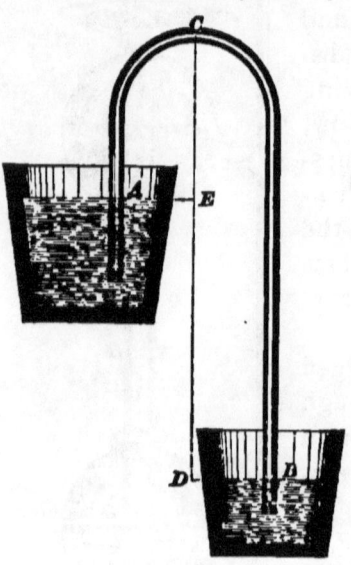

FIG. 126.

will be greater than that upon the longer column, and this excess pressure will produce motion.

605. Let A = the area of the tube;

$h = DC$ = the vertical distance between the surface of the water in B, and the highest point of the center line of the tube;

$h_1 = EC$ = the distance between the surface of the water in A, and the highest point of the center line of the tube.

The weight of the water in the short column is $.03617 \times A\, h_1$, and the resultant atmospheric pressure tending to force the water up the short column is $14.7 \times A - .03617 \times A\, h_1$. The weight of the water in the long column is $.03617\, A\, h$, and the resultant atmospheric pressure tending to force the water up the long column is $14.7\, A - .03617\, A\, h$. The difference between these two is $(14.7\, A - .03617\, A\, h_1) - (14.7\, A - .03617\, A\, h) = .03617\, A\, (h - h_1)$. But $h - h_1 = ED$ = the difference between the levels of the water in the two vessels. In the above, h and h_1 were taken in inches, and A in square inches.

It will be noticed that the short column must not be higher than 34 feet for water, or the siphon will not work, since the pressure of the atmosphere will not support a column of water that is higher than 34 feet.

606. The Injector.—A section of an injector is shown in Fig. 127. There are many different kinds of these instruments, but the principle is the same in all. They are used for lifting water from a point below the discharge orifice, and forcing it into the boiler of a steam engine or locomotive. They depend for their action upon the creation of a partial vacuum by the action of steam. The valve A is opened by turning the hand wheel S, and the steam enters from F into D, and flows through the tubes C, K, M, and the outlet N, into the boiler. The valve B is opened by turning the hand wheel W. The steam, flowing through C with a high velocity, drives out the air. The air in the

240 MECHANICS.

chamber GG and in the suction pipe P expands and flows into C through the conical orifice H, thus creating a partial vacuum in the chamber G and pipe P. The atmospheric

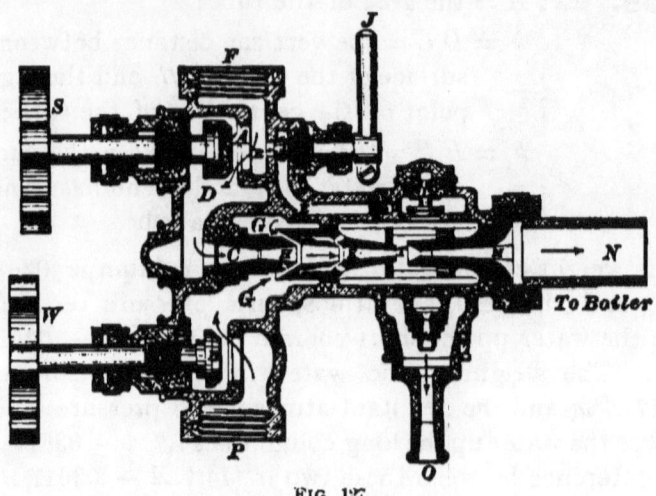

FIG. 127.

pressure on the water causes it to rise in pipe P, and flow into G, and from thence into C, through the conical orifice H, whence the steam drives it through K, M, and N into the boiler.

607. The method of operating the injector is as follows: The water valve B is opened; next the small valve R is opened by turning the handle J. The steam then flows into the passage E, which connects with the chamber D through the conical tube C, and creates a partial vacuum in G, as before described. The quantity of steam admitted by the valve R not being sufficient to force all of the water which flows through the orifice H into the boiler, the water will accumulate in the chamber T, raise the valve L, and flow down through u and out at the overflow outlet O. As soon as the water appears at O the valve A is opened as far as possible, and the valve R closed. If the water still flows out at O close the valve B slightly until it stops. An injector will lift water, according to temperature and steam pressure, from 6 to 26 feet.

PUMPS.

608. The Suction Pump.—A section of an ordinary suction pump is shown in Fig. 128. Suppose the piston to be at the bottom of the cylinder, and to be just on the point of moving upwards in the direction of the arrow. As the piston rises it leaves a vacuum behind it, and the atmospheric pressure upon the surface of the water in the well causes it to rise in the pipe P for the same reason that the mercury rises in the barometer tube. The water rushes up the pipe and lifts the valve V, filling the empty space in the cylinder B displaced by the piston. When the piston has reached the end of its stroke the water entirely fills the space between the bottom of the piston and the bottom of the cylinder, and also the pipe P. The instant that the piston begins its down stroke, the water in the chamber B tends to fall back into the well, and its weight forces the valve V to its seat, thus preventing any downward flow of the water. As the piston descends the water must give way to it, and, since the valve V is closed, the valves u, u must open, and thus allow the water to pass through the piston, as shown in the right-hand figure. When the piston has reached the end of its downward stroke, the weight of the water above closes the valves u, u. All the water resting on the top of the piston is then lifted with the piston on its upward stroke, and discharged through the spout A, the valve V again opening, and the water filling the space below the piston as before.

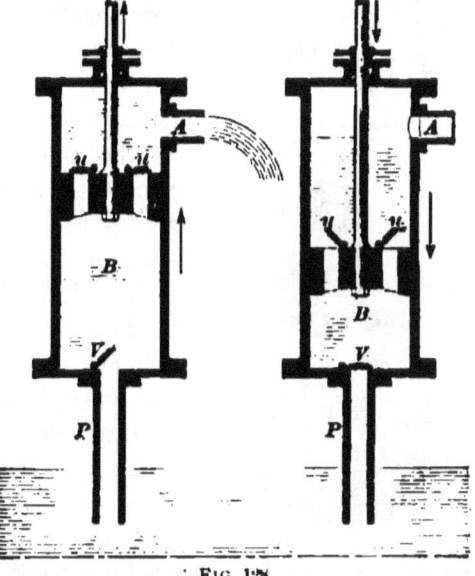

Fig. 128.

609. It is evident that the distance between the piston when at top of its stroke and the surface of the water in the well must not exceed 34 feet, the highest column of water which the pressure of the atmosphere will sustain, since, otherwise, the water in the pipe would not rise up and fill the cylinder as the piston ascended. In practice, this distance should not exceed 28 feet. This is due to the fact that there is a little air left between the bottom of the piston and the bottom of the cylinder, a little air leaks through the valves, which are not perfectly air-tight, and a pressure is needed to raise the valve against its weight, which, of course, acts downwards. There are many varieties of the suction pump, differing principally in the valves and piston, but the principle is the same in all.

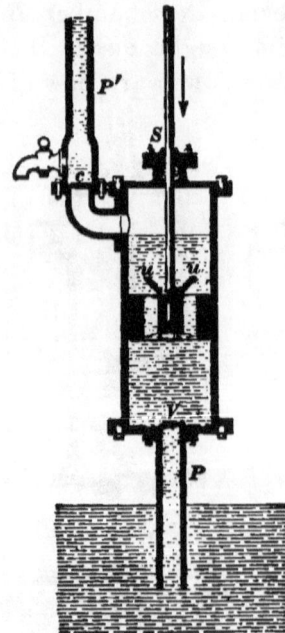

Fig. 129.

610. The Lifting Pump.— A section of a lifting pump is shown in Fig. 129. These pumps are used when water is to be raised to greater heights than can be done with the ordinary suction pump. As will be perceived, it is essentially the same as the pump previously described, except that the spout is fitted with a cock and has a pipe attached to it, leading to the point of discharge. If it is desired to discharge the water at the spout, the cock may be opened; otherwise, the cock is closed, and the water is lifted by the piston up through the pipe P to the point of discharge, the valve C preventing it from falling back into the pump, and the valve V preventing the water in the pump from falling back into the well.

611. Force Pumps.—The force pump differs from the lifting pump in several important particulars, but chiefly in the fact that the piston is solid; that is, it has no valves. A section of a suction and force pump is shown in Fig. 130. The water is drawn up the suction pipe as before, when the piston rises; but when the piston reverses, the pressure on the water caused by the descent of the piston opens the valve V' and *forces* the water up the delivery pipe P'. When the piston again begins its upward movement, the valve V' is closed by the pressure of the water above it, and the valve V is opened by the pressure of the atmosphere on the water below it, as in the previous cases. For an arrangement of this kind, it is not necessary to have a stuffing-box. The water may be forced to almost any desired height. The force pump differs again from the lifting pump in respect to its piston rod, which should not be longer than is absolutely necessary in order to prevent it from *buckling*, while in the lifting pump the length of the piston rod is a matter of indifference.

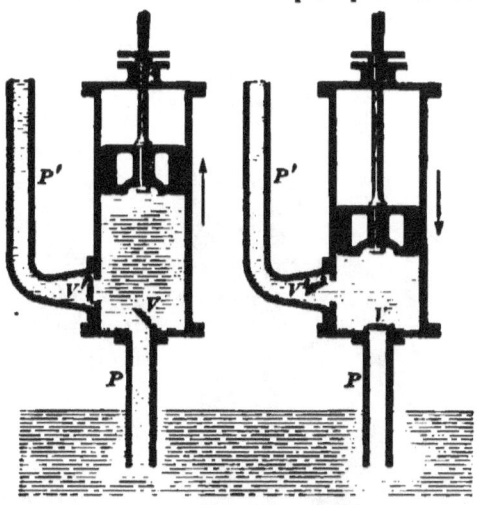

Fig. 130.

612. Plunger Pumps.—When force pumps are used to convey water to great heights, the pressure of the water in the cylinder becomes so great that it becomes extremely difficult to keep the water from leaking past the piston, and the constant repairing of the piston packing becomes a nuisance. To obviate this difficulty the piston is made very long, as shown in Fig. 131, and is then called the **plunger**. The suction valve in this case consists of two clack valves inclined to each other and resting upon a square pin A; they are prevented from flying back too far during the up stroke of the plunger by the two uprights I, I. During the down stroke of the plunger, the valves at *A* are closed and the valve *B* in the delivery pipe is open. A little air is always carried into the cylinder of a pump with the entering water. In force pumps this fact becomes a serious consideration, since after repeated strokes the air accumulates, and during the down stroke of the plunger it is compressed. After a time it would become sufficient to entirely prevent the water from entering through the suction valve, the pressure on the top of the valve being greater than that of the atmosphere below. In the pump shown in the figure, the plunger is a trifle smaller than the cylinder, and the air collects around the plunger below the stuffing-box. To remove this air a narrow passage *C* (shown by the dotted lines), that can be closed at its upper end by the cock *D*, connects the interior of the pump with the atmosphere when the cock is open. It is evident that this cock must not be opened except during the down stroke of the plunger, for, if it were open during the up stroke, the pressure below the plunger being less than the pressure of the atmosphere above, the air would rush in instead of being expelled.

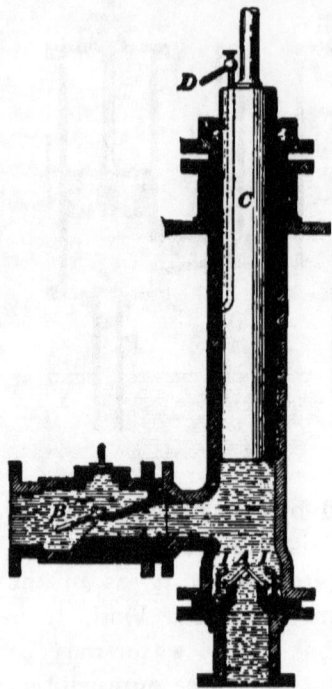

FIG. 131.

AIR CHAMBERS.

613. In order to obtain a continuous flow of water in the delivery pipe, with as nearly a uniform velocity as possible, an **air chamber** is usually placed on the delivery pipe of force pumps as near to the pump cylinder as the construction of the machine will allow. The air chambers are usually pear-shaped, with the small end connected to the

MECHANICS. 245

pipe. They are filled with air, which the water compresses during the discharge. During the suction, the air thus compressed expands and acts as an accelerating force upon the moving column of water, a force which diminishes with the expansion of the air and helps to keep the velocity of the moving column more nearly uniform. An air chamber is sometimes placed upon the suction pipe. These air chambers not only tend to promote a uniform discharge, but also to equalize the stresses upon the pump and prevent shocks due to the incompressibility of water. They subserve the same purpose on pumps that a fly-wheel does on the steam engine. Unless the pump moves very slowly, it is absolutely necessary to have an air chamber on the delivery pipe.

STEAM PUMPS.

614. Steam pumps are force pumps operated by steam acting upon the piston of a steam engine directly connected to the pump, and in many cases cast with the pump. A

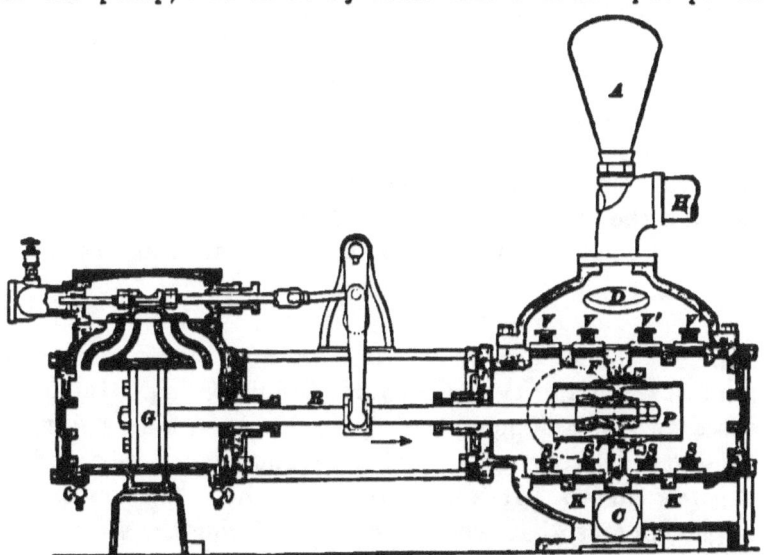

FIG. 132.

section of a double-acting steam pump, showing the steam and water cylinders, with other details, is illustrated in Fig. 132. Here G is the steam piston, and R the piston rod, which is secured at its other end to the pump plunger P. F is a partition cast with the cylinder, which prevents the water in the left-hand half from communicating with that in the right-hand half of the cylinder. Suppose the piston to be moving in the direction of the arrow. When the piston has arrived at the end of its stroke, the water space in the left-hand half of the pump cylinder will have been increased by an amount equal to the area of the cross-section of the plunger, multiplied by the length of the stroke, and the volume of the right-hand half of the cylinder will have been diminished by a like amount. In consequence of this, a volume of water in the right-hand half of the cylinder equal to the volume displaced by the plunger in its forward movement will be forced through the valves $V,' V'$, through the orifice D, into the air chamber A, and then discharged through the delivery pipe H. By reason of the partial vacuum in the left-hand half of the pump cylinder, owing to this movement of the plunger, the water will be drawn from the reservoir through the suction pipe C into the chamber $K K$, lifting the valves S', S' and filling the space displaced by the plunger. During the return stroke the water will be drawn through the valves S, S into the right-hand half of the pump cylinder, and at the same time water will be discharged from the left-hand half through the valves V, $V,'$ out through the pipe H, as before. Each one of the four suction and four discharge valves is kept to its seat, when not working, by light springs, as shown.

615. There are many varieties and makes of steam pumps, the majority of which are double-acting. In many cases two steam pumps are placed side by side, having a common delivery pipe. This arrangement is called a **duplex pump.** It is usual to so set the steam pistons of duplex pumps that when one is completing the stroke, the other is in the middle of its stroke. A double-acting duplex pump made to run in this manner, and having an air chamber of sufficient size, will deliver water with nearly a uniform velocity.

MECHANICS. 247

In mine pumps for forcing water to great heights, the plungers are made solid, and in most cases extend through the pump cylinder. In many steam pumps, pistons are used instead of plungers, but when very heavy duty is required, plungers are preferred.

STRENGTH OF MATERIALS.

616. When a force is applied to a body, it changes either its form or its volume. A force, when considered with reference to the internal changes it tends to produce in any solid, is called a **stress**.

Thus, if we suspend a weight of 2 tons by a rod, the stress in the rod is 2 tons. This stress is accompanied by a lengthening of the rod, which increases until the internal stress or resistance is in equilibrium with the external weight.

617. Stresses may be classified as follows:

Tensile, or pulling stress. | Transverse, or bending stress.
Compressive, or pushing stress. | Shearing, or cutting stress.
 | Torsional, or twisting stress.

618. A **unit stress** is the amount of stress on a unit of area, and may be expressed either in pounds per square inch, or in tons per square foot; or, it is the load per square inch or square foot on any body.

Thus, if 10 tons are suspended by a wrought-iron bar which has an area of 5 square inches, the unit stress is 2 tons per square inch, because $\dfrac{10}{5} = 2$ tons.

619. **Strain** is the deformation or change of shape of a body resulting from stress.

For example, if a rod 100 feet long is pulled in the direction of its length, and if it is lengthened 1 foot, it is strained $\frac{1}{100}$th of its length, or 1 per cent.

620. **Elasticity** is the power which a body has of returning to its original form after the external force on it is withdrawn, providing the stress has not exceeded the elastic limit.

Consequently, we see from this that all material is lengthened or shortened when subjected to either tensile or compressive stress, and the change of the length is directly proportional to the stress, within the elastic limit.

For stresses within the elastic limits, materials are perfectly elastic, and return to their original length on removal of the stresses; but, when their elastic limits are exceeded, the changes of their lengths are no longer regular, and a permanent **set** takes place; the destruction of the material has then begun.

621. The **measure of elasticity** of any material is the change of length under stress within the elastic limit.

622. The **elastic limit** is that unit stress under which the permanent set becomes visible.

The elasticity of wrought iron is practically the same as that of steel; that is, each material will change an equal amount of length under the same stress within the elastic limits.

The elastic limit of steel is higher than that of wrought iron; consequently, the former will lengthen or shorten more than the latter before its elasticity is injured.

TENSILE STRENGTH OF MATERIALS.

623. The tensile strength of any material is the resistance offered by its fibers to being pulled apart.

The tensile strength of any material is proportional to the area of its cross-section.

Consequently, when it is required to find the safe tensile strength of any material, we have only to find the area at the minimum cross-section of the body, and multiply it by its strength per square inch, as given in the following table under the heading "Working Stress."

NOTE.—The minimum cross-section referred to in the above paragraph is that section of the material which is pierced with holes; such as bolt or rivet holes in iron, or knots in wood, if there are any.

624. In the following table is given the average breaking and working tensile stress of different materials:

TABLE 17.

Material.	Breaking Stress in Pounds per Square Inch.	Working Stress in Pounds per Square Inch.
Timber............	10,000	600 to 1,200
Cast Iron.........	16,000	1,500 to 3,500
Wrought Iron.....	50,000	5,000 to 12,000
Steel.............	70,000	6,000 to 13,000

The above table shows that the tensile breaking strength of cast iron is 16,000 pounds per square inch of cross-section, and that the working strength is from 1,500 to 3,500 pounds per square inch of cross-section.

625. In machinery, such as steam engines, where the parts are subjected to shocks, or are alternately compressed and extended, it is not safe to subject cast iron to a stress of more than 1,500 pounds per square inch of section, wrought iron to more than 5,000 pounds per square inch of section, or steel to more than 6,000 pounds per square inch of section.

But in structures in which the strains are constantly in one direction, as is the case with steam boilers, wrought iron may be strained with from 6,000 to 8,000 pounds per square inch of section, or steel with from 8,000 to 10,000 pounds per square inch of section.

Consequently, strict attention must be given to the *nature of the load* the given structure has to bear, and fix the working stress accordingly.

NOTE.—For structures on which the load is applied suddenly, use the smaller working stresses given in the table, and for those on which the load is applied gradually, use the larger working stresses.

RULES AND FORMULAS FOR TENSILE STRENGTH.

626. Let W = safe load in pounds;

A = area of minimum cross-section;

S = working stress in pounds per square inch, as given in the foregoing table.

Rule 111.—*The working load in pounds for any bar subjected to a tensile stress is equal to the minimum sectional area of the bar, multiplied by the working stress in pounds per square inch, as given in the table.*

That is, $W = A\,S$.

EXAMPLE.—A bar of good wrought iron which is 3″ square is to be subjected to a steady tensile stress; what is the maximum load that it should carry?

SOLUTION.—From what has been said above in regard to the materials and to the nature of the load, it will be safe in this case to use a working stress of 12,000 pounds per square inch.

Applying the rule, we have

$$W = 3 \times 3 \times 12{,}000 = 108{,}000 \text{ pounds. Ans.}$$

Rule 112.—*The minimum sectional area of any bar subjected to a tensile stress should be equal to the load in pounds, divided by the working stress in pounds per square inch, as given in the table.*

That is, $A = \dfrac{W}{S}$.

EXAMPLE.—What should be the area of a wrought-iron bar to carry a steady load of 108,000 pounds, if it is to resist a tensile stress of 12,000 pounds per square inch?

SOLUTION.—Applying the rule,

$$A = \frac{108{,}000}{12{,}000} = 9 \text{ sq. in. Ans.}$$

Rule 113.—*The working stress in pounds per square inch is equal to the load in pounds divided by the minimum sectional area of the bar.*

That is, $S = \dfrac{W}{A}$.

MECHANICS.

EXAMPLE.—A bar of wrought iron 3" square, subjected to tensile stress, carries a load of 108,000 pounds; what is the stress per square inch?

SOLUTION.—Applying the rule,

$$S = \frac{108,000}{3 \times 3} = 12,000 \text{ lb. per sq. in. Ans.}$$

CHAINS.

627. Chains made of the same size iron vary in strength, owing to the different kinds of links from which they are made.

It is a good practice to anneal old chains which have become brittle by overstraining. This renders them less liable to snap from sudden jerks. It reduces their tensile strength, but increases their toughness and ductility, which are sometimes more important qualities.

When annealing, care should be taken that a sufficient heat be applied, otherwise no benefit will be gained; the chains ought to be heated to a cherry red, say 1300° F. at the least.

628. Rules and Formulas for the Strength of Chains:

Let W = safe load in pounds;

D = diameter of the iron, in inches, from which the links are made.

Rule 114.—*The safe load in pounds of a stud-link wrought-iron chain is equal to 18,000, multiplied by the square of the diameter of the iron from which the links are made.*

That is, $W = 18,000 \, D^2$.

EXAMPLE.—What is the maximum load that should be carried by a stud-link wrought-iron chain, if its links are made from ¾-inch round iron?

SOLUTION.—Applying the rule, $W = 18,000 \, D^2$.

Substituting the value of D^2, we have $W = 18,000 \times \frac{3}{4} \times \frac{3}{4} = 10,125$ pounds. Ans.

Rule 115.—*The safe load in pounds of a close-link wrought-iron chain is equal to 12,000, multiplied by the square of the diameter of the iron from which the links are made.*

That is, $W = 12{,}000\ D^2$.

EXAMPLE.—What is the maximum load that should be carried by a close-link wrought-iron chain, if its links are made from $\tfrac{3}{4}$-inch round iron?

SOLUTION.—Applying the rule, $W = 12{,}000\ D^2$.

Substituting the value of D^2, we have $W = 12{,}000 \times \dfrac{3}{4} \times \dfrac{3}{4} = 6{,}750$ pounds. Ans.

HEMP ROPES.

629. The strength of hemp ropes does not depend so much upon the quality of the material and the cross-section of the rope, as upon the method of manufacture and the amount of twisting.

The ropes in common use are three-strand, shroud-laid rope, and hawser or cable-laid rope.

The strongest ropes are three-strand shroud-laid, made without tar. Ropes made with tar are less flexible, and are reduced in strength about 25 per cent., but have better wearing qualities.

630. Rules and Formulas for the Strength of Hemp Ropes:

Let W = maximum working load in pounds;

C = circumference of rope in inches.

Rule 116.—*The maximum working load in pounds that should be allowed on any hemp rope is equal to the square of the circumference of the rope, multiplied by 100.*

That is, $W = 100\ C^2$.

EXAMPLE.—What is the maximum load in pounds that should be carried by a hemp rope which has a circumference of 8 inches?

SOLUTION.—Substituting the value of C in the formula, $W = 100 \times 8^2 = 6{,}400$ lb. Ans.

Rule 117.—*The circumference of any hemp rope is equal to the square root of the maximum working load in pounds which it is capable of carrying, multiplied by .1.*

That is, $C = .1 \sqrt{W}$.

EXAMPLE.—A maximum working load of 1,000 pounds is to be carried by a hemp rope; what should be the circumference of the rope?

SOLUTION.—Applying the rule, $C = .1 \sqrt{1,000} = 3.16''$. Ans.

When measuring ropes, the circumference is sought instead of the diameter, because the ropes are not round and the circumference therefore is not 3.1416 times the diameter. For three strands the circumference is about $2.86\, d$; for seven strands, $3\, d$.

631. The above formulas are very convenient for use, and easily remembered, but it is well to remark that the values thus given apply to ropes of a *good average* quality.

WIRE ROPES.

632. Wire rope is made of iron and steel wire. It is stronger than hemp rope, and, to carry the same load, is of smaller diameter.

In substituting steel for iron rope, the object in view should be to gain an increase of wear from the rope, rather than to reduce the size.

A steel rope to be serviceable should be of the best obtainable quality, because ropes made from low grades of steel are inferior to good iron ropes.

633. Formulas for the Strength of Wire Ropes:
Let W = maximum of working load in pounds;
C = circumference of rope in inches.

Rule 118.—*The maximum working load in pounds that should be allowed on any iron wire rope is equal to the square of the circumference of the rope in inches, multiplied by 600.*

That is, $W = 600\, C^2$.

EXAMPLE.—What is the maximum load in pounds that should be carried by an iron wire rope whose circumference is 4½ inches?

SOLUTION.—Applying the formula,
$W = 600 \times 4.5^2 = 12{,}150$ lb. Ans.

MECHANICS.

Rule 119.—*The circumference of any iron wire rope in inches is equal to the square root of the maximum working load in pounds, multiplied by .0408.*

That is, $C = .0408 \sqrt{W}$.

EXAMPLE.—A maximum working load of 12,150 pounds is to be carried by an iron wire rope; what should be the minimum circumference of the rope?

SOLUTION.—Applying the formula,

$C = .0408 \sqrt{12,150} = 4\frac{1}{2}$ inches. Ans.

Rule 120.—*The above rules and formulas are also made applicable when computing the safe strength of steel wire rope, by substituting the constant 1,000 for the constant 600, and .0316 for .0408.*

EXAMPLE.—What is the maximum load in pounds that should be carried by a steel wire rope, the circumference of which is 4½ inches?
SOLUTION.—Applying the rule, $W = 1,000 \times 4.5^2 = 20,250$ lb. Ans.

EXAMPLE.—A maximum working load of 10,485 pounds is to be carried by a steel wire rope; what should be the minimum circumference of the rope?
SOLUTION.—Applying the rule, $C = .0316 \sqrt{10,485} = 3.24$ inches.

Ans.

EXAMPLES FOR PRACTICE.

1. What should be the diameter of a steel piston rod of a steam engine to resist tension, if the piston is 19″ in diameter and the pressure is 85 lb. per sq. in.? Ans. 2¼″, nearly.

2. What safe load will a cast-iron bar of rectangular cross-section 7½″ by 3¼″ support if subjected to shocks? The bar is in tension.
Ans. 39,375 lb.

3. What is the stress per sq. in. on a piece of timber 8″ square, which is subjected to a steady pull of 60,000 pounds?
Ans. 937.5 lb. per sq. in.

4. What should be the safe load for a close-link wrought-iron chain whose links are made from ⅞″ iron? Ans. 9,187.5 lb.

5. What safe load may a hemp rope carry whose circumference is 4″? Ans. 1,600 lb.

6. What should be the allowable working load for a steel wire rope whose circumference is 3¾″? Ans. 14,062.5 lb.

7. What should be the circumference of an iron wire rope to support a load of 20,000 lb.? Ans. 5¾″, nearly.

MECHANICS.

CRUSHING STRENGTH OF MATERIALS.

634. The crushing strength of any material is the resistance offered by its fibers to being pushed together.

To obtain only compression, the length of a rod should not be more than 5 times greater than its least diameter, or its least thickness when it is a rectangular rod.

If a bar is long compared with its cross dimensions, the load if sufficiently great will cause it to bend sideways under the compressive force, and we have, then, not only compression, but compression compounded with bending.

Experimental tests on pillars have shown that their strengths are approximately inversely proportional to the squares of their lengths. That is, if there are two pillars of the same material, having the same cross-section, but one is twice as long as the other, the long one will sustain only about one-quarter the load of the short one.

635. Attention should be given to the ends of pillars, as their shape has great influence upon their strength. In Fig. 133 are shown three pillars with differently shaped ends

FIG 133.

It has been proven by the aid of higher mathematics that, theoretically, a pillar having flat or fixed ends, as shown at a, is four times as strong as one that has round or movable ends, as shown at c, and one and seven-ninths times as strong as one having one flat and one round end, as shown at b; b is thus two and one-fourth times as strong as c. It has also been found that if three pillars, a, b, c, which have the same cross-section, are to carry the same load and be of equal strength, their lengths must be as the numbers 2, 1½, and 1, respectively.

In practice, however, the ends of the pillars b and c are not generally made as shown by the figure, but have holes at their ends into which pins are fitted which are fastened to some other piece; as, for example, the connecting-rod of an engine. In such cases, it has been found that a is two times as strong as c, and that b is one and one-half times as strong as c. That is, in actual practice, a column fixed as at c is really $\frac{1}{2}$ as strong as one fixed as at a, instead of being only $\frac{1}{4}$ as strong, as given above.

Green or wet timber has only one-half the strength of dry and seasoned timber; consequently, its crushing strength is only one-half of that given in the table below.

636. In the following table is given the mean crushing strength of some short specimens of materials in tons (of 2,000 pounds) per square inch:

TABLE 18.

Materials.	Crushing Strength in Tons per Square Inch.
Cast Iron	40
Wrought Iron	18
Mild Steel	26
Cast Copper	5
Cast Brass	4.5
Timber (dry) endwise	3.5
Brick	1
Stone	3

637. Formula for the Strength of Pillars:

The following formula is applicable to pillars which are commonly used in practice, the lengths of which are about from 10 to 40 times their least diameter, or, if rectangular, their least thickness as indicated by d.

Let $C =$ crushing strength in tons per square inch of a short specimen of the material, as given in the above table;

$S =$ sectional area in square inches;

$L =$ length in inches;

MECHANICS. 257

$d =$ least thickness of rectangular pillar, or diameter of round pillar in inches;
$W =$ breaking load in tons;
$A =$ the area of the two flanges;
$B =$ the area of the web;
$a =$ constant, given in one of the following three tables:

TABLE 19.
CONSTANTS FOR WROUGHT-IRON PILLARS.

Cross-section of Pillar.	When Both Ends of the Pillar are Flat or Fixed.	When One End of the Pillar is Flat or Fixed, and the Other Round or Movable.	When Both Ends of the Pillar are Round or Movable.
Round.	2,250	1,500	1,125
Square or Rectangle.	3,000	2,000	1,500
Thin Square Tube.	6,000	4,000	3,000
Thin Round Tube.	4,500	3,000	2,250
Angle with Equal Sides.	1,500	1,000	750
Cross with Equal Arms	1,500	1,000	750
I Beam.	$3{,}000 \times \dfrac{A}{A+B}$	$2{,}000 \times \dfrac{A}{A+B}$	$1{,}500 \times \dfrac{A}{A+B}$

MECHANICS.

TABLE 20.
CONSTANTS FOR CAST-IRON PILLARS.

Cross-section of Pillar.	When Both Ends of the Pillar are Flat or Fixed.	When One End of the Pillar is Flat or Fixed, and the Other Round or Movable.	When Both Ends of the Pillar are Round or Movable.
Round.	281.25	187.5	140.625
Square or Rectangle.	375.	250.	187.5
Thin Square Tube.	750.	500.	375.
Thin Round Tube.	562.5	375.	281.25
Angle with Equal Sides.	187.5	125.	93.75
Cross with Equal Arms.	187.5	125.	93.75
I Beam.	$375 \times \dfrac{A}{A+B}$	$250 \times \dfrac{A}{A+B}$	$125 \times \dfrac{A}{A+B}$

TABLE 21.
CONSTANTS FOR WOODEN PILLARS.

Cross-section of Pillar.	When Both Ends of the Pillar are Flat or Fixed.	When One End of the Pillar is Flat or Fixed, and the Other Round or Movable.	When Both Ends of the Pillar are Round or Movable.
Round.	187.5	125.	93.75
Square or Rectangle.	250.	166.66	125.
Hollow Square Made of Boards.	500.	333.33	250.

638. Rule 121.—*The breaking load of a pillar in tons is equal to the crushing strength of a short specimen of the material as given in Table 18, Art.* **636**, *multiplied by the sectional area of the pillar in square inches, and the product divided by 1 plus the quotient obtained by dividing the square of the length of the pillar in inches by the square of the diameter (or least thickness, if rectangular) multiplied by the value of a.*

That is,
$$W = \frac{CS}{1 + \frac{L^2}{a\,d^2}}$$

The result obtained by the formula must be divided by 6 to get the safe working load.

NOTE.—If the length of the pillar is given in feet, be sure to reduce it to inches before substituting in the formula.

EXAMPLE.—A wooden pillar, 6 inches square and 144 inches long, is fixed at both ends; what load will it sustain with safety?

SOLUTION.—By rule **121**, $W = \dfrac{CS}{1 + \frac{L^2}{a\,d^2}}$.

Substituting the values of C, S, L^2, a, and d^2, we have

$$W = \frac{3.5 \times 6 \times 6}{1 + \frac{144 \times 144}{250 \times 6 \times 6}} = 38.14 \text{ tons, nearly.}$$

Which, divided by 6, gives $\frac{38.14}{6} = 6.357$ tons, or the load it is capable of sustaining with safety. Ans.

EXAMPLE.—A wrought-iron pillar, 4 inches in diameter and 60 inches long, is fixed at one end and movable at the other; what load will it sustain with safety?

SOLUTION.—By the rule,
$$W = \frac{18 \times 4 \times 4 \times .7854}{1 + \frac{60 \times 60}{1{,}500 \times 4 \times 4}} = 196.69 \text{ tons.}$$

Which, divided by 6, gives $\frac{196.69}{6} = 32.78$ tons, nearly, or the load it is capable of sustaining with safety. Ans.

EXAMPLE.—What load will a cast-iron pillar sustain with safety, if it is 20 feet long, and if its cross-section is a cross with equal arms, two of which are equal to 10 inches in length (as d, see table), and whose arms are 1 inch thick; both ends of the pillar movable.

SOLUTION.—Area of cross-section $= (10 \times 1) + 2(4.5 \times 1) = 19$ square inches; 20 feet $= 240$ inches.
$$W = \frac{40 \times 19}{1 + \frac{240 \times 240}{93.75 \times 10 \times 10}} = 106.38 \text{ tons.}$$

Which, divided by 6, gives $\frac{106.38}{6} = 17.73$ tons, the load it is capable of sustaining with safety. Ans.

639. When using this formula, first obtain the value of C from Table 18, Art. **636.** Next, calculate the area of the cross-section of the pillar. Then, find the value of a from one of the last four tables. Finally, be sure that the length of the pillar has been reduced to inches before substituting in the formula.

To find the proper value of a in any example, first turn to the table dealing with the material in question, and find the figure corresponding to the given cross-section; in the horizontal line containing this are three numbers corresponding to the different conditions of the ends of the column. From these numbers select the one corresponding to the given conditions of the column to be calculated, and this will be the required value of a.

MECHANICS.

EXAMPLES FOR PRACTICE.

1. What load may be safely carried by a hollow cylindrical cast-iron pillar, 20 ft. long, inside diameter 8", and outside diameter 10"? Both ends of the pillar are fixed. Ans. 93.13 tons.

2. A rectangular wooden column is 14 ft. long, and has one end rounded; if the cross-section is 12" × 8", what load will be required to break it? Ans. 92.15 tons.

3. A solid wrought-iron column, which has both ends movable, is 3" in diameter and 8 ft. long; what load will it safely support? Ans. 11.1 tons.

TRANSVERSE STRENGTH OF MATERIALS.

640. The transverse strength of any material is the resistance offered by its fibers to being broken by bending. As, for example, when a beam, bar, rod, etc., which is supported at its ends, is broken by a force applied between its supports.

The transverse strength of any beam, bar, rod, etc., is proportional to the product of the square of its depth multiplied by its width; consequently, it is more economical to increase the depth than the width.

TABLE 22.
CONSTANTS FOR TRANSVERSE STRENGTH.

Material.	Constant in Pounds.	Material.	Constant in Pounds.
Metals:		Woods:	
Cast Iron	100	Birch	35
Wrought Iron	150	Elm	25
Structural Steel	160	Ash	45
Copper	50	Beech	30
Brass	55	Hickory	50
		Maple	60
		Oak (American)	45
		Pine (Pitch)	40
		Pine (White)	30

641. A **cantilever** is a beam, bar, rod, etc., fixed at one end and subjected to a transverse stress, as shown in

Fig. 134. It has a tendency to overthrow the wall or structure to which it is attached.

The strength of a cantilever varies inversely as the distance of the load from the point of fixing; and the stress upon any section varies directly as the distance of the load from that section.

The strength of a beam, bar, rod, etc., which has both its ends supported, but not fixed, and which carries a load midway between its supports, is four times that of a beam of the same length, fixed into a wall at one end, and carrying a load at the other, as in Fig. 134.

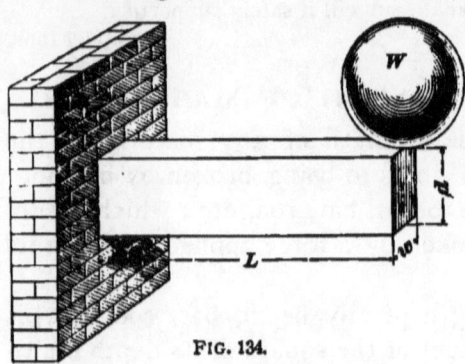

FIG. 134.

A cantilever uniformly loaded will sustain twice as great a load as one in which the load is applied at the free end; and a beam resting on two supports and uniformly loaded will sustain twice as great a load as it would if the load were all applied at the middle.

In Table 22 is given the safe transverse strength of bars of different kinds of material, one inch square and one foot long, with the load suspended from one end, the other being fixed, as shown in Fig. 134.

642. Rules and Formulas for the Transverse Strength of Beams:

Let $d=$ the depth of beam in inches;
$w=$ the width of the beam in inches;
$L=$ the length of the beam between its supports, in feet, or, for cantilever, the distance between load and fixed end;
$S=$ the safe transverse strength, as given in Table 22;
$W=$ the safe load in pounds.

MECHANICS.

For a rectangular or square cantilever to which the load is applied at one end, as shown in Fig. 134:

Rule 122.—*The maximum safe load in pounds that should be allowed at the end of any rectangular or square cantilever is equal to the square of the depth in inches multiplied by its width in inches multiplied by the constant given in the table, and the product divided by its length in feet.*

That is, $W = \dfrac{d^2 w S}{L}$.

EXAMPLE.—What is the maximum safe load that can be placed at one end of a cast-iron bar which projects 4 feet, the depth being 6 inches, and the width 3 inches?

SOLUTION.—By the rule, $W = \dfrac{d^2 w S}{L}$.

Substituting the values of d, w, S, and L, we have

$$W = \frac{6 \times 6 \times 3 \times 100}{4} = 2,700 \text{ pounds. Ans.}$$

643. For a cylindrical cantilever to which the load is applied at one end, as shown in Fig. 135:

Rule 123.—*The maximum safe load in pounds that should be allowed at the end of any cylindrical cantilever is equal to the cube of its diameter in inches multiplied by .6 of the constant given in the table, and the product divided by its length in feet.*

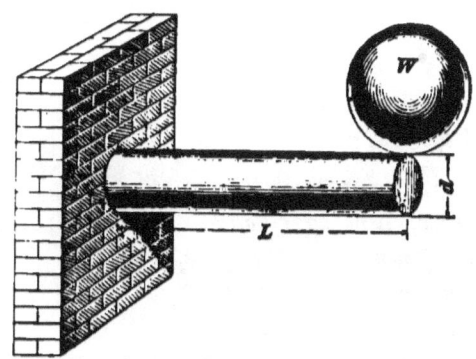

FIG. 135.

That is, $W = \dfrac{d^3 \times .6 S}{L}$.

EXAMPLE.—What is the maximum load that can be placed with safety at one end of a cast-iron bar 4 inches in diameter that projects 3 feet?

SOLUTION.—By the rule, $W = \dfrac{d^3 \times .6S}{L}$.

Substituting the values of d, S, and L, we have

$$W = \frac{4 \times 4 \times 4 \times .6 \times 100}{3} = 1{,}280 \text{ pounds. Ans.}$$

644. When the load is uniformly distributed on a cantilever of any cross-section, as shown in Fig. 136, it will sustain a load twice as great as when the load is applied at one end. For example, if the cantilevers in the two examples above were to carry a uniformly distributed load, they would sustain $2{,}700 \times 2 = 5{,}400$ pounds, and $1{,}280 \times 2 = 2{,}560$ pounds respectively.

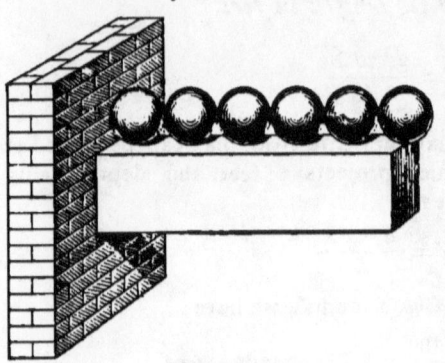

FIG. 136.

For a rectangular or square beam the ends of which merely rest upon supports, and loaded in the middle, as shown in Fig. 137:

Rule 124.—*The maximum safe load in pounds that any rectangular or square beam is capable of sustaining at the middle, when its ends merely rest upon supports, is equal to four times the square of its depth in inches multiplied by its width in inches multiplied by the constant given in the table, and the product divided by the distance between its supports in feet.*

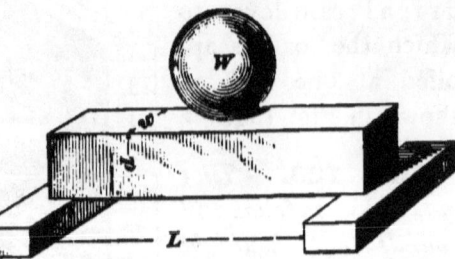

FIG. 137.

That is, $\qquad W = \dfrac{4\, d^2\, w\, S}{L}.$

EXAMPLE.—What maximum safe load is a bar of cast iron capable of sustaining in the middle between the supports on which its ends merely rest, if its depth is 6 inches, its width 3 inches, and the distance between the supports is 4 feet?

SOLUTION.—By the rule, $W = \dfrac{4 \times 6^2 \times 3 \times 100}{4} = 10{,}800$ lb. Ans.

645. For a cylindrical beam supported at its ends and loaded in the middle, as shown in Fig. 138:

Rule 125. — *The maximum safe load in pounds that any cylindrical beam is capable of sustaining at the middle, when*

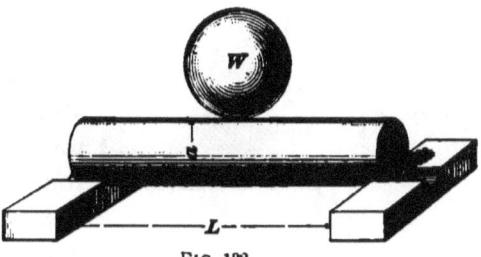

FIG. 138.

its ends merely rest upon supports, is equal to four times the cube of its diameter multiplied by .6 of the constant given in the table, and the product divided by the distance between its supports in feet.

That is, $W = \dfrac{4\, d^3 \times .6\, S}{L}$.

EXAMPLE.—What maximum safe load is a bar of cast iron capable of sustaining in the middle between the supports on which its ends merely rest, if it is 4 inches in diameter, and if the distance between its supports is 3 feet?

SOLUTION.—By the rule, $W = \dfrac{4 \times 4^3 \times .6 \times 100}{3} = 5{,}120$ lb. Ans.

646. When the load is uniformly distributed on a beam of any cross-section, as shown in Fig. 139, it will sustain a load twice as great as when the load is applied in the middle between the supports.

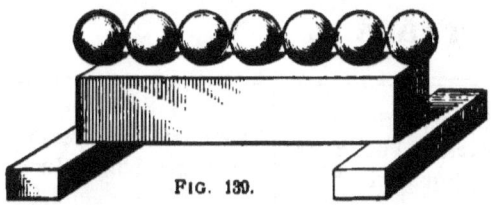

FIG. 139.

For example, if the beams in the last two examples were to carry a uniformly distributed load, they would sustain $10{,}800 \times 2 = 21{,}600$ pounds, and $5{,}120 \times 2 = 10{,}240$ pounds, respectively.

SHEARING, OR CUTTING, STRENGTH OF MATERIALS.

647. The shearing strength of any material is the resistance offered by its fibers to being cut in two.

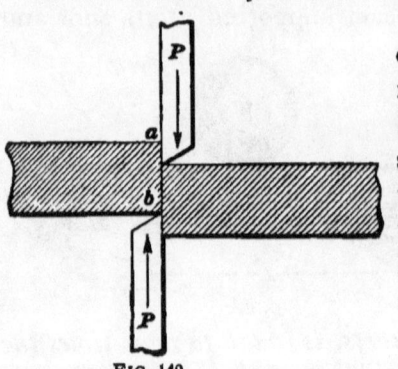

Thus, the pressure of the cutting edges of an ordinary shearing machine, Fig. 140, causes a shearing stress in the plane $a\,b$. The unit shearing force may be found by dividing the force P by the area of the plane $a\,b$.

FIG. 140.

648. Fig. 141 shows a piece in **double shear**; here the central piece $c\,d$ is forced out while the ends remain on their supports M and N.

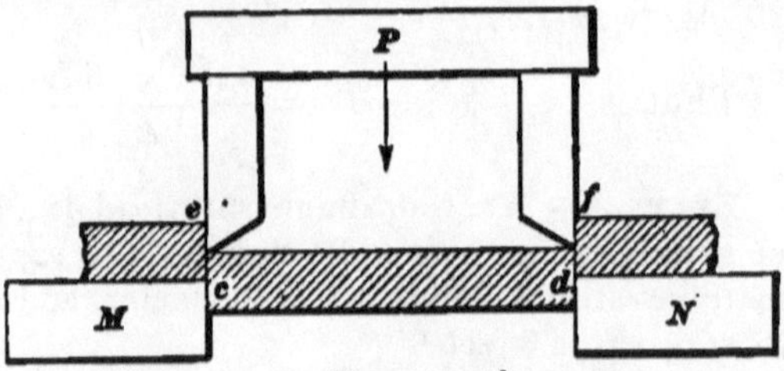

The shearing strength of any body is directly proportional to its area.

FIG. 141.

In the following table are given the greatest and safe shearing strengths per square inch of different kinds of materials:

TABLE 23.

Material.	Greatest Shearing Stress in Pounds per Square Inch.	Safe Shearing Stress in Pounds per Square Inch.
Cast Iron.........	18,000	1,500 to 3,000
Wrought Iron.....	40,000	4,000 to 10,000
Steel	60,000	5,000 to 12,000

MECHANICS.

649. Formula for the Shearing Strength of Materials:

Let a = area of cross-section in inches;
S = safe shearing stress as given in the table;
W = safe load in pounds.

Rule 126.—*The safe load that any body which is subjected to a shearing stress is capable of sustaining is equal to the area of its cross-section in inches multiplied by its safe shearing stress, as given in the table.*

That is, $W = a\,S$.

EXAMPLE.—If the beam in Fig. 141 is made of wrought iron, 4 inches in depth and 2 inches in width, what *steady* shearing stress is it capable of sustaining with safety?

SOLUTION.—Applying the rule, $W = 4 \times 2 \times 10,000 = 80,000$ lb. This result must be multiplied by 2, since the beam is sheared in two places, along the lines ec and fd. Hence, the stress which the beam will safely sustain is $80,000 \times 2 = 160,000$ lb.

EXAMPLE.—What force is required to punch a hole $\frac{5}{8}''$ in diameter through a steel plate $\frac{1}{2}''$ thick?

SOLUTION.—It is evident that punching is but shearing in a circle instead of a straight line. The area punched (sheared) is equal to the thickness of the plate multiplied by the circumference of a circle having the same diameter as the punched hole. For, if the plate were cut through one of the diameters of the punched hole, and the two semicircles were straightened out, the punched surface would be a rectangle, which would have a length equal to the circumference of a circle whose diameter was equal to that of the hole, and a breadth equal to the thickness of the plate. In this case, the area $= \frac{5}{8} \times 3.1416 \times \frac{1}{2} = .98175$ sq. in. Table 23 gives the ultimate shearing strength of steel as 60,000 lb. per sq. in. Hence, the total force required is $.98175 \times 60,000 = 58,905$ lb. Ans.

EXAMPLES FOR PRACTICE.

1. What is the greatest load that can be safely carried by a steel rectangular cantilever at its extreme end, if the bar is 2" wide, 3" deep, and 2 ft. 6" long? Ans. 1,152 lb.

2. What is the greatest uniform load that can be safely carried by a white-pine girder, 6" wide, 8" deep, 16 ft. long, and supported at its ends? Ans. 5,760 lb.

MECHANICS.

3. A cast-iron bar, $1\frac{3}{4}''$ in diameter and 5 ft. 3" long, is supported at its ends; what load will it safely sustain in the middle? Ans. 245 lb.

4. What force is required to punch a $1\frac{1}{4}''$ hole through a wrought-iron plate $\frac{7}{16}''$ thick? Ans. 68,723 lb.

5. What force is required to cut off the end of a cast-iron bar whose diameter is $2\frac{1}{4}''$? Ans. 88,357 lb.

LINE SHAFTINGS.

650. A line of shafting is one continuous run, or length, composed of lengths of shafts joined together by couplings.

The **main line** of shafting is that which receives the power from the engine or motor, and distributes it to the other lines of shafting or to the various machines to be driven.

Line shafting is supported by hangers, which are brackets provided with bearings, bolted either to the walls, posts, ceilings, or floors of the building. Short lengths of shafting, called **countershafts,** are provided to effect changes of speed, and to enable the machinery to be stopped or started.

Shafting is usually made cylindrically true, either by a special rolling process, when it is known as **cold-rolled shafting,** or else it is turned up in a machine called a lathe. In the latter case it is called **bright shafting.** What is known as **black shafting** is simply bar iron rolled by the ordinary process, and turned where it receives the couplings, pulleys, bearings, etc.

Bright turned shafting varies in diameter by $\frac{1}{4}$ inch up to about $3\frac{1}{2}$ inches in diameter; above this diameter the shafting varies by $\frac{1}{2}$ inch. The actual diameter of a bright shaft is $\frac{1}{16}$ of an inch less than the commercial diameter, it being designated from the diameter of the ordinary round bar iron from which it is turned. Thus, a length of what is called 3-inch bright shafting is only $2\frac{15}{16}$ inches in diameter.

Cold-rolled shafting is designated by its commercial diameter; thus, a length of what is called 3-inch shafting is 3 inches in diameter.

651. In the following table is given the maximum distance between the bearings of some continuous shafts which are used for the transmission of power:

TABLE 24.

Diameter of Shaft in Inches.	Distance Between Bearings in Feet.	
	Wrought-Iron Shaft.	Steel Shaft.
2	11	11.5
3	13	13.75
4	15	15.75
5	17	18.25
6	19	20.
7	21	22.25
8	23	24.
9	25	26.

Pulleys from which considerable power is to be taken should always be placed as close to a bearing as possible.

652. The diameters of the different lengths of shafts composing a line of shafting may be proportional to the quantity of power delivered by each respective length. In this connection, the positions of the various pulleys depend upon the distance between the pulley and the bearing, and upon the amount of power given off by the pulleys. Suppose, for example, that a piece of shafting delivers a certain amount of power, then, it is obvious that the shaft will deflect or bend less if the pulley transmitting that power be placed close to a hanger or bearing, than if it be placed midway between the two hangers or bearings. It is impossible to give any rule for the proper distance of bearings which could be used universally, as in some cases the requirements demand that the bearings be nearer together than in others.

If the work done by a line of shafting is distributed quite equally along its entire length, and the power can be applied near the middle, the strength of the shaft need be only half as great as would be required if the power were applied at one end.

653. To compute the horsepower that can be transmitted by a shaft of any given diameter:

Let D = diameter of shaft;
R = revolutions per minute;
H = horsepower transmitted;
C = constant given in Table 25.

TABLE 25.
CONSTANTS FOR LINE SHAFTING.

Material of Shaft.	No Pulleys Between Bearings.	Pulleys Between Bearings.
Steel or Cold-Rolled Iron	65	85
Wrought Iron..........	70	95
Cast Iron.............	90	120

In the above table the bearings are supposed to be spaced so as to relieve the shaft of excessive bending; also, in the third vertical column, an average number and weight of pulleys, and power given off, is assumed.

In determining the above constants, allowance has been made to insure the stiffness as well as strength of the shaft. Cold-rolled iron is considerably stronger than ordinary turned wrought iron; the increased strength being due to the process of rolling, which seems to compress the metal and so make it denser, not merely skin deep, but practically throughout the whole diameter. We have, then, the following:

Rule 127.—*The horsepower that a shaft will transmit equals the product of the cube of the diameter and the number of revolutions, divided by the value of C for the given material.*

That is, $$H = \frac{D^3 \times R}{C}.$$

MECHANICS.

EXAMPLE.—What horsepower will a 3-inch wrought-iron shaft transmit which makes 100 revolutions per minute, there being no pulleys between bearings?

SOLUTION.—$H = \dfrac{D^3 \times R}{C}$.

Substituting, we have

$$H = \frac{3 \times 3 \times 3 \times 100}{70} = 38.57 \text{ horsepower. Ans.}$$

If there were the usual amount of power taken off, as mentioned above, we should take $C = 95$. Then, $H = \dfrac{27 \times 100}{95} = 28.42$ horsepower. Ans.

654. To compute the number of revolutions a shaft must make to transmit a given horsepower:

Rule 128.—*The number of revolutions necessary for a given horsepower equals the product of the value of C for the given material and the number of horsepower, divided by the cube of the diameter.*

That is, $\qquad R = \dfrac{C \times H}{D^3}$.

EXAMPLE.—How many revolutions must a 3-inch wrought-iron shaft make per minute to transmit 28.42 horsepower, power being taken off at intervals between the bearings?

SOLUTION.—$R = \dfrac{C \times H}{D^3}$.

Substituting, we have $R = \dfrac{95 \times 28.42}{3 \times 3 \times 3} = 100$ revolutions. Ans.

655. To compute the diameter of a shaft that will transmit a given horsepower, the number of revolutions the shaft makes per minute being given:

Rule 129.—*The diameter of a shaft equals the cube root of the quotient obtained by dividing the product of the value of C for the given material and the number of horsepower by the number of revolutions.*

That is, $\qquad D = \sqrt[3]{\dfrac{C \times H}{R}}$.

EXAMPLE.—What must be the diameter of a wrought-iron shaft to transmit 38.57 horsepower, the shaft to make 100 revolutions per minute, no power being taken off between bearings?

SOLUTION.—$D = \sqrt[3]{\dfrac{C \times H}{R}}$.

Substituting, we have

$$D = \sqrt[3]{\frac{70 \times 38.57}{100}} = \sqrt[3]{27} = 3 \text{ in. Ans.}$$

MECHANICS.

As the speed of shafting is used as a multiplier in the calculations of the horsepower of shafts, a shaft having a given diameter will transmit more power in proportion as its speed is increased. Thus, a shaft which is capable of transmitting 10 horsepower when making 100 revolutions per minute, will transmit 20 horsepower when making 200 revolutions per minute. We may, therefore, say *the horsepowers transmitted by two shafts are directly proportional to the number of revolutions.*

EXAMPLES FOR PRACTICE.

1. What horsepower will a $2\frac{1}{4}''$ wrought-iron shaft transmit when running at 110 revolutions per minute, it being used for transmission only? Ans. 24.55 horsepower.

2. A $6''$ cast-iron shaft transmits 150 horsepower; how many revolutions per minute must it make, no power being taken off between bearings? Ans. $62\frac{1}{2}$ R. P. M.

3. What should be the diameter of a wrought-iron shaft to transmit 100 horsepower at 150 revolutions per minute, power being taken off between bearings? Ans 4 in., nearly.

4. The diameter of a steam-engine shaft is $8''$; what horsepower will it transmit, if made of steel, when making 150 revolutions per minute? Ans. 1,181.54 horsepower.

5. The machines driven by a certain line of wrought-iron shafting take their power from various points between the bearings; and, if all were working together at their full capacity, they would require 65 horsepower to drive them. What diameter should the shaft be if it runs at 150 revolutions per minute? Ans. $3\frac{1}{2}$ in.

A SERIES

OF

QUESTIONS AND EXAMPLES

RELATING TO THE SUBJECTS
TREATED OF IN VOL. I.

It will be noticed that, although the various questions are numbered in sequence from **1** to **383,** inclusive, these questions are divided into five different sections, corresponding to the five sections of the preceding pages of this volume. Under the heading of each section are given, in parentheses, the numbers of those articles which should be carefully studied before attempting to answer any question or to solve any example occurring in the section.

ARITHMETIC.

(SEE ARTS. 1 to 174.)

(1) What is arithmetic ?

(2) What is a number ?

(3) What is the difference between a concrete number and an abstract number ?

(4) Define notation and numeration.

(5) Write each of the following numbers in words: 980; 605; 28,284; 9,006,042; 850,317,002; 700,004.

(6) Represent in figures the following expressions:
Seven thousand, six hundred. Eighty-one thousand, four hundred two. Five million, four thousand, seven. One hundred eight million, ten thousand, one. Eighteen million, six. Thirty thousand, ten.

(7) What is a fraction ?

(8) What are the terms of a fraction ?

(9) What does the denominator show ?

(10) What does the numerator show ?

(11) How do you find the value of a fraction ?

(12) Is $\frac{13}{8}$ a proper or an improper fraction and why ?

(13) Write three mixed numbers.

(14) Reduce the following fractions to their lowest terms: $\frac{4}{8}, \frac{4}{16}, \frac{8}{32}, \frac{11}{44}$.   Ans. $\frac{1}{2}, \frac{1}{4}, \frac{1}{4}, \frac{1}{4}$.

(15) Reduce 6 to an improper fraction whose denominator is 4.   Ans. $\frac{24}{4}$.

(16) Reduce $7\frac{1}{8}$, $13\frac{5}{16}$, and $10\frac{3}{4}$ to improper fractions.
Ans. $\frac{57}{8}, \frac{213}{16}, \frac{43}{4}$.

and 17 thousandths, and the sum of 53 hundredths and 274 thousandths?

Ans. $\begin{cases} (a) & .068. \\ (b) & .5. \\ (c) & .45125. \\ (d) & .786967. \end{cases}$

(42) It is desired to increase the capacity of an electric-light plant to 1,500 horsepower by adding a new engine. If the indicated horsepower of the engines already in use is 482¼, 316⅓, and 390¾, what power must the new engine develop? Ans. 310.11⅔ H. P.

(43) On a certain morning 7,240 gallons of water were drawn from an engine-room tank, and 4,780 gallons were pumped in. In the afternoon 7,633 gallons were drawn out, and 8,675 gallons pumped in. How many gallons remained in the tank at night, if it contained 3,040 gallons at the beginning of the day? Ans. 1,622 gal.

(44) A metal stack 45 feet high is made up of 7 plates, 6 of which are 7 feet long. Allowing 15 inches for laps, what is the length of the seventh plate? Ans. 4 ft. 3 in.

(45) The inside diameter of a 6-inch steam pipe is 6.06 inches, and the outside diameter is 6.62 inches. How thick is the pipe? Ans. .28 in.

(46) Find the products of the following:
(a) 526,387 × 7; (b) 700,298 × 17; (c) 217 × 103 × 67.

Ans. $\begin{cases} (a) & 3,684,709. \\ (b) & 11,905,066. \\ (c) & 1,497,517. \end{cases}$

(47) If your watch ticks once every second, how many times will it tick in one week? Ans. 604,800.

(48) Find the products of the following expressions:
(a) .013 × .107; (b) 203 × 2.03 × .203; (c) (2.7 × 31.85) × (3.16 − .316); (d) (107.8 + 6.541 − 31.96) × 1.742.

Ans. $\begin{cases} (a) & .001391. \\ (b) & 83.65427. \\ (c) & 244.56978. \\ (d) & 143.507702. \end{cases}$

ARITHMETIC. 279

(49) An engine and boiler in a manufactory are worth $3,246. The building is worth three times as much plus $1,200, and the tools are worth twice as much as the building plus $1,875; (*a*) what is the value of the building and tools? (*b*) What is the value of the whole plant?

Ans. $\begin{cases} (a)\ \$34,689. \\ (b)\ \$37,935. \end{cases}$

(50) How many square feet of heating surface are in the tubes of a boiler having 60 3-inch tubes, each $15\frac{1}{2}$ feet long, if the heating surface of each tube per foot in length is .728 square foot? Ans. 677.04 sq. ft.

(51) Suppose that in one hour 10 pounds of coal are burned per square foot of grate area in a certain boiler, and that 9 pounds of water are evaporated per pound of coal burned. If the grate area is 30 square feet, how many pounds of water would be evaporated in a day of 10 hours?

Ans. 27,000 lb.

(52) How many feet does the piston of a steam engine pass over in a week of 6 days, running $8\frac{1}{2}$ hours a day, if the length of the stroke of the engine is $1\frac{1}{2}$ feet, and the number of revolutions per minute 160? Ans. 1,468,800 ft.

(53) A number of boilers are constantly fed by 3 pumps; the first delivers $1\frac{1}{4}$ gallons per stroke, and runs at 75 strokes per minute; the second delivers $\frac{7}{8}$ of a gallon per stroke, and runs at 115 strokes per minute; the third delivers $1\frac{3}{8}$ gallons per stroke, and runs at 96 strokes per minute. How many gallons of water are fed to the boilers per hour?

Ans. 21,022.5 gal.

NOTE.—If in the following examples there be a remainder, carry the quotient to four decimal places.

(54) Divide the following:

(*a*) 962,842 by 84; (*b*) 39,728 by 63; (*c*) 29,714 by 108; (*d*) 406,089 by 135.

Ans. $\begin{cases} (a)\ 11,462.4048. \\ (b)\ 630.6032. \\ (c)\ 275.1296. \\ (d)\ 3,008.0667. \end{cases}$

(55) Solve the following:

(a) $35 \div \frac{5}{16}$; (b) $\frac{9}{16} \div 3$; (c) $\frac{17}{2} \div 9$; (d) $\frac{113}{64} \div \frac{7}{16}$, (e) $15\frac{3}{4} \div 4\frac{3}{8}$.

Ans. $\begin{cases} (a) & 112. \\ (b) & \frac{3}{16}. \\ (c) & \frac{17}{18}. \\ (d) & 4\frac{1}{28}. \\ (e) & 3\frac{3}{5}. \end{cases}$

(56) Solve the following:

(a) $.875 \div \frac{1}{2}$; (b) $\frac{7}{8} \div .5$; (c) $\dfrac{.375 \times \frac{1}{4}}{\frac{5}{16} - .125}$. Ans. $\begin{cases} (a) & 1.75. \\ (b) & 1.75. \\ (c) & .5. \end{cases}$

(57) Solve the following by cancelation:

(a) $(72 \times 48 \times 28 \times 5) \div (84 \times 15 \times 7 \times 6)$.

(b) $(80 \times 60 \times 50 \times 16 \times 14) \div (70 \times 50 \times 24 \times 20)$.

Ans. $\begin{cases} (a) & 9\frac{1}{7}. \\ (b) & 32. \end{cases}$

(58) Find the values of the following expressions:

(a) $\dfrac{7}{\frac{3}{16}}$; (b) $\dfrac{1\frac{1}{8}}{\frac{3}{8}}$; (c) $\dfrac{1.25 \times 20 \times 3}{\dfrac{87 + 88}{459 + 32}}$. Ans. $\begin{cases} (a) & 37\frac{1}{3}. \\ (b) & .75. \\ (c) & 210\frac{3}{4}. \end{cases}$

(59) The distance around a cylindrical boiler is 166.85 inches. If there are 72 rivets in one of the circular seams, find what the pitch (distance between the centers of any two rivets) of the rivets is. Ans. $2.317 +$ in.

(60) A keg of $\frac{7}{8} \times 2\frac{3}{4}$ inches boiler rivets weighs 100 pounds, and contains 133 rivets. What is the weight of each rivet? Ans. $.75 +$ lb.

(61) The distance around a wheel equals 3.1416 times its diameter, or the distance across it. If the distance around a fly-wheel is 56.5488 feet, what is the diameter of a wheel half as large? Ans. 9 ft.

(62) If 980,000 bricks are required to build an engine house, how many days will it take for 6 teams to draw them, each team drawing 8 loads a day, and there being 1,600 bricks to a load? Ans. $12.76 +$ days.

ARITHMETIC.

(63) If a mechanic earns $1,500 a year for his labor, and his expenses are $968 per year, in what time can he save enough to buy 28 acres of land at $133 an acre?
Ans. 7 years.

(64) The numerator of a fraction is 28, and the value of the fraction $\frac{7}{8}$; what is the denominator?   Ans. 32.

(65) From 1 plus .001 take .01 plus .000001.
Ans. .990999.

(66) A freight train ran 365 miles in one week, and 3 times as far, lacking 246 miles, the next week; how far did it run the second week?   Ans. 849 miles.

(67) If the driving wheel of a locomotive is 16 ft. in circumference, how many revolutions will it make in going from Philadelphia to Pittsburg, the distance of which is 354 miles, there being 5,280 feet in one mile?
Ans. 116,820 rev.

(68) How many inches in .875 of a foot?   Ans. 10½ in.

(69) What decimal part of a foot is $\frac{3}{16}$ of an inch?
Ans. .015625 ft.

(70) If water be conducted from a tank by 7 lengths of gas pipe coupled together, each length being 12 feet 6 inches long, ¾ of an inch being added at each of the joints for coupling together, how far from the tank is the water discharged? The first length screws into the tank ¾ of an inch.   Ans. 87.8125 ft.

(71) If by selling a carload of coal for $82.50, at a profit of $1.65 per ton, I make enough to pay for 72.6 ft. of fencing at $.50 a foot, how many tons of coal were there in the car?   Ans. 22 tons.

(72) The connection between an engine and boiler is made up of 6 lengths of pipe, three of which are 14 feet 5 inches long, two 12 feet 6 inches long, and one 8 feet 10 inches long. If the pipe weighs 10½ pounds per foot, what is the total weight of the pipe used?   Ans. 809.375 lb.

(73) Four bolts are required, 2½, 6¼, 3$\frac{1}{16}$, and 4 inches long. How long a piece of iron will be required from

which to cut them, allowing $\frac{7}{16}$ of an inch to each bolt for cutting off and finishing? Ans. $18\frac{3}{16}$.

(74) A double belt of a certain width can transmit $64\frac{1}{2}$ horsepower. How many horsepower can two single belts of the same width transmit, when running under the same conditions, supposing that the double belt is capable of transmitting $\frac{10}{7}$ as much power as one of the single belts?

Ans. 90.3 H. P.

(75) The lengths of belting required to connect four countershafts with the main line shaft were 18 feet 6 inches, 16 feet $9\frac{1}{2}$ inches, 22 feet 2 inches, and 20 feet $8\frac{1}{4}$ inches. How many feet of belting were required? Ans. $78\frac{1}{4}$ ft.

(76) In a steam-engine test of an hour's duration, the horsepower developed was found to be as follows, at 10-minute intervals: 48.63, 45.7, 46.32, 47.9, 48.74, 48.38, 48.59. What was the average horsepower?

Ans. $47.75 +$ H. P.

ARITHMETIC.
(SEE ARTS. 175–339.)

(77) What is 25% of 8,428 lb. ?   Ans. 2,107 lb.
(78) What is 1% of $100 ?   Ans. $1.
(79) What is ½% of $35,000 ?   Ans. $175.
(80) What per cent. of 50 is 2 ?   Ans. 4%.
(81) What per cent. of 10 is 10 ?   Ans. 100%.
(82) Solve the following:

(*a*) Base = $2,522, and percentage = $176.54. What is the rate?

(*b*) Percentage = 16.96, and rate = 8%. What is the base?

(*c*) Amount = 216.7025, and base = 213.5. What is the rate?

(*d*) Difference = 201.825, and base = 207. What is the rate?

Ans. $\begin{cases} (a) \ 7\%. \\ (b) \ 212. \\ (c) \ 1\tfrac{1}{2}\%. \\ (d) \ 2\tfrac{1}{2}\%. \end{cases}$

(83) The coal consumption of a steam plant is 5,500 lb. per day when the condenser is not running, or an increase of 15% over the consumption when the condenser is used. How many pounds are used per day when the condenser is running?   Ans. 4,782.61 lb.

(84) An engineer receives a salary of $950. He pays 24% of it for board, 12½% of it for clothing, and 17% of it for other expenses. How much does he save a year?   Ans. $441.75.

(85) If 37½% of a number is 961.38, what is the number?   Ans. 2,563.68.

(86) A man owns ¾ of a manufacturing plant. 30% of his share is worth $1,125. What is the whole property worth?
Ans. $5,000.

(87) What number diminished by 35% of itself equals 4,810? Ans. 7,400.

(88) The volume of the clearance in a steam-engine cylinder is found to be 18.3 cu. in., and the volume of the cylinder, neglecting the clearance, 254.5 cu. in. What percentage of the cylinder volume is the clearance?
Ans. 7.2%, nearly.

(89) The distance between two stations on a certain railroad is 16.5 miles, which is 12½% of the entire length of the road. What is the length of the road? Ans. 132 miles.

(90) The speed of an engine running unloaded was 1½% greater than when running loaded. If it made 298 revolutions per minute with the load, what was its speed running unloaded? Ans. 302.47 revolutions per minute.

(91) Reduce 4 yd. 2 ft. 10 in. to inches. Ans. 178 in.

(92) Reduce 3,722 in. to higher denominations.
Ans. 103 yd. 1 ft. 2 in.

(93) How many seconds in 5 weeks and 3.5 days?
Ans. 3,326,400 sec.

(94) Reduce 764,325 cu. in. to cu. yd.
Ans. 16 cu. yd. 10 cu. ft. 549 cu. in.

(95) How many gallons of water can be put into a tank holding 4 bbl. 10 gal. 3 qt.? Ans. 136¾ gal.

(96) A carload of coal weighed 16 T. 8 cwt. 75 lb. How many pounds did this amount to? Ans. 32,875 lb.

(97) Reduce 25,396 lb. to higher denominations.
Ans. 12 T. 13 cwt. 96 lb.

(98) Reduce 25,396 pt. to higher denominations.
Ans. 100 bbl. 24 gal. 2 qt.

(99) What is the sum of 2 yd. 2 ft. 3 in.; 4 yd. 1 ft. 9 in.; 2 ft. 7 in.? Ans. 8 yd. 7 in.

ARITHMETIC.

(100) What is the sum of 3 gal. 3 qt. 1 pt.; 6 gal. 1 pt.; 4 gal. 8 qt. 5 pt.? Ans. 16 gal. 2 qt. 1 pt.

(101) From 52 yd. 2 ft. 9 in. take 115 ft.
Ans. 14 yd. 1 ft. 9 in.

(102) From a barrel of machine oil is sold at one time 10 gal. 2 qt. 1 pt., and at another time 16 gal. 3 qt. How much remained? Ans. 4 gal. 1 pt.

(103) If 1 iron rail is 17 ft. 3 in. long, how long would 51 such rails be, if placed end to end? Ans. 879 ft. 9 in.

(104) Multiply 3 qt. 1 pt. by 4.7. Ans. 32.9 pt.

(105) How many iron rails, each 30 ft. long, are required to lay a single railroad track 23 miles long?
Ans. 8,096 rails.

(106) The main line shaft of a mill is composed of 4 lengths each 15 ft. 5 in. long, of one piece 14 ft. 8 in. long, and one piece 8 ft. 10 in. long. If there are 6 hangers spaced equally distant apart, one being placed at each extremity of the shaft, what is the distance between the hangers? Ans. 17 ft. ⅘ in.

(107) The distance around a wheel is approximately $\frac{22}{7}$ times the diameter. If the diameter of a fly-wheel is 9 ft. 6¼ in., find the distance around it. Ans. 29 ft. 11⅞ in.

(108) If the length of a boiler shell is 18 ft. 11¼ in., how many rivets should there be in one of the longitudinal seams if it is a single-riveted seam, supposing the rivets to have a pitch of 1¼ in., and the two end rivets to be 1¼ in. from each end of the boiler? Ans. 181 rivets.

(109)

(a) What is the second power of 108?

(b) What is the third power of 181.25?

(c) What is the fourth power of 27.61?

Ans. $\begin{cases} (a)\ 11,664. \\ (b)\ 5,954,345.703125. \\ (c)\ 581,119.73780641. \end{cases}$

(110) Solve the following:

(a) 106^2; (b) $(182\frac{1}{8})^2$; (c) $.005^2$; (d) $.0063^2$; (e) 10.06^2; (f) 67.85^2; (g) $967,845^2$; (h) $(\frac{7}{18})^2$; (i) $(\frac{1}{4})^2$.

Ans.
- (a) 11,236.
- (b) 33,169.515625.
- (c) .000025.
- (d) .00003969.
- (e) 101.2036.
- (f) 4,603.6225.
- (g) 936,723,944,025.
- (h) $\frac{49}{256}$.
- (i) $\frac{1}{16}$.

(111) Solve the following:

(a) 753^2; (b) 987.4^2; (c) $.005^3$; (d) $.4044^2$; (e) $.0133^2$; (f) 301.011^2; (g) $(\frac{1}{8})^3$; (h) $(3\frac{3}{4})^3$.

Ans.
- (a) 426,957,777.
- (b) 962,674,279.624.
- (c) .000000125.
- (d) .066135317184.
- (e) .000002352637.
- (f) 27,273,890.942264331.
- (g) $\frac{1}{512}$.
- (h) 52.734375.

(112) What is the fifth power of 2? Ans. 32.

(113) What is the fourth power of 3? Ans. 81.

(114) What is the ninth power of 7? Ans. 40,353,607.

(115) Solve the following:

(a) 1.2^4; (b) 11^5; (c) 1^5; (d) $.01^4$; (e) $.1^5$.

Ans.
- (a) 2.0736.
- (b) 161,051.
- (c) 1.
- (d) .00000001.
- (e) .00001.

(116) In what respect does evolution differ from involution?

ARITHMETIC

(117) Find the square root of the following:

(a) 3,486,784.401; (b) 9,000,099.4009; (c) .001225; (d) 10,795.21; (e) 73,008.05; (f) 9; (g) .9.

Ans. $\begin{cases} (a) & 1,867.29+. \\ (b) & 3,000.016. \\ (c) & .035. \\ (d) & 103.9. \\ (e) & 270.2. \\ (f) & 3. \\ (g) & .948+. \end{cases}$

(118) What is the cube root of the following:

(a) .32768? (b) 74,088? (c) 92,416? (d) .373248? (e) 1,758.416743? (f) 1,191,016? (g) $\frac{4}{32}$? (h) $\frac{27}{256}$?

Ans. $\begin{cases} (a) & .689+. \\ (b) & 42. \\ (c) & 45.2115+. \\ (d) & .72. \\ (e) & 12.07. \\ (f) & 106. \\ (g) & .5. \\ (h) & .472+. \end{cases}$

(119) Find the cube root of 2 to four decimal places.

Ans. $1.2599+$.

(120) Find the cube root of 3 to three decimal places.

Ans. $1.442+$.

(121) Solve the following:

(a) $\sqrt{123.21}$;   (b) $\sqrt{114.921}$;
(c) $\sqrt{502,681}$;   (d) $\sqrt{.00041209}$.

Ans. $\begin{cases} (a) & 11.1. \\ (b) & 10.72+. \\ (c) & 709. \\ (d) & .0203. \end{cases}$

(122) Solve the following:

(a) $\sqrt[3]{.0065}$; (b) $\sqrt[3]{.021}$; (c) $\sqrt[3]{8,036,054,027}$;
(d) $\sqrt[3]{.000004096}$; (e) $\sqrt[3]{17}$.

Ans. $\begin{cases} (a)\ .186+. \\ (b)\ .27+. \\ (c)\ 2,003. \\ (d)\ .016. \\ (e)\ 2.57+. \end{cases}$

(123) Extract the square root of:

(a) $1\frac{281}{4176}$.
(b) .3364.
(c) .1.
(d) $25.0\frac{3}{4}$.
(e) $.000\frac{1}{3}$.

Ans. $\begin{cases} (a)\ \frac{31}{44}. \\ (b)\ .58. \\ (c)\ .31622+. \\ (d)\ 5.00749. \\ (e)\ .02108. \end{cases}$

(124) Find the value of x in the following:

11.7 : 13 :: 20 : x. Ans. 22.22 +.

(125) (a) $20 + 7 : 10 + 8 :: 3 : x$.
(b) $12^2 : 100^2 :: 4 : x$.

Ans. $\begin{cases} (a)\ 2. \\ (b)\ 277.7+. \end{cases}$

(126) (a) $\frac{4}{x} = \frac{7}{21}$; (b) $\frac{x}{24} = \frac{8}{16}$; (c) $\frac{2}{10} = \frac{x}{100}$; (d) $\frac{15}{45} = \frac{60}{x}$;

(e) $\frac{10}{150} = \frac{x}{600}$.

Ans. $\begin{cases} (a)\ x = 12. \\ (b)\ x = 12. \\ (c)\ x = 20. \\ (d)\ x = 180. \\ (e)\ x = 40. \end{cases}$

(127) $x : 5 :: 27 : 12.5$. Ans. $10\frac{4}{5}$.

(128) $.45 : 60 :: x : 24$. Ans. 18.

(129) $x : 35 :: 4 : 7$. Ans. 20.

(130) $9 : x :: 6 : 24$. Ans. 36.

(131) $\sqrt[3]{1,000} : \sqrt[3]{1,331} :: 27 : x$. Ans. 29.7.

(132) $64 : 81 :: 21^2 : x^2$. Ans. 23.625.

(133) $7 + 8 : 7 :: 30 : x$. Ans. 14.

(134) If a piece of 2-inch shafting $3\frac{1}{2}$ ft. long weighs 37.45 lb., how much would a piece $6\frac{3}{4}$ ft. long weigh?

Ans. 72.225 lb.

ARITHMETIC.

(135) The intensity of heat from a burning body varies inversely as the square of the distance from it. If a thermometer held 6 ft. from a stove rises 24 degrees, how many degrees will it rise if held 12 ft. from the stove ? Ans. 6°.

(136) If sound travels at the rate of 6,160 ft. in $5\frac{1}{2}$ sec., how far does it travel in 1 min.? Ans. 67,200 ft.

(137) If a railway train runs 444 mi. in 8 hr. 40 min., in what time can it run 1,060 mi. at the same rate of speed ? Ans. 20 hr. 41.44 min.

(138) If a pump discharging 135 gal. per minute fills a tank in 38 minutes, how long would it take a pump discharging 85 gal. per minute to fill it ? Ans. $60\frac{6}{17}$ min.

(139) If a quantity of babbitt metal contains 8 lb. of copper, 8 lb. of antimony, and 80 lb. of tin, how much copper will 32 lb. of the same metal contain ? Ans. $2\frac{2}{3}$ lb.

(140) The distances around the drive wheels of two locomotives are 12.56 ft. and 15.7 ft., respectively. How many times will the larger turn while the smaller turns 520 times ? Ans. 416 times.

(141) If a cistern 28 ft. long, 12 ft. wide, 10 ft. deep holds 510 bbl. of water, how many barrels of water will a cistern hold that is 20 ft. long, 17 ft. wide, and 6 ft. deep ? Ans. $309\frac{9}{14}$ bbl.

MENSURATION AND USE OF LETTERS

IN

ALGEBRAIC FORMULAS.

(SEE ARTS. 340–432.)

$A = 5 \qquad h = 200$

$B = 10 \qquad x = 12$

$i = 3.5 \qquad D = 120$

Work out the solutions to the following formulas, using the above values for the letters:

(142) $C = \dfrac{D - x}{B + i}.$ \qquad Ans. $C = 8.$

(143) $Q = \dfrac{A h + D}{2 x + 6} + D.$ \qquad Ans. $Q = 157\frac{1}{3}.$

(144) $r = \dfrac{3.246\, B h}{\dfrac{A x + h}{A i - B}}.$ \qquad Ans. $r = 187.269 +.$

(145) $v = \sqrt{\dfrac{A D}{i B + 1.5}}.$ \qquad Ans. $v = 4.05 +.$

(146) $u = \sqrt[3]{\dfrac{B x}{.00018\, h (A^2 - x)}}.$ \qquad Ans. $u = 6.35 +.$

(147) $f = \dfrac{10 (h - D)^2}{\sqrt[3]{D + A}}.$ \qquad Ans. $f = 12{,}800.$

(148) $g = \dfrac{(B - A)^2 - \sqrt[3]{D + A}}{A^2 - (1 + D)}.$ \qquad Ans. $g = 5.$

(149) $k = \sqrt{\dfrac{A B^2}{\sqrt[3]{A h}}}.$ \qquad Ans. $k = 7.071 +.$

(150) $T = \sqrt{\dfrac{A^2\left[490 + \dfrac{(hx)^2}{D^2}\right]}{h + \dfrac{x}{D}(A^2 - B)^2}}$. Ans. $T = 10$.

(151) If one of the angles formed by one straight line meeting another straight line equals 152° 3′, what is the other angle equal to? Ans. 27° 57′.

(152) How many seconds are there in 140° 17′ 10″.
Ans. 505,030″.

(153) Write a definition for a degree, as applied to the measurement of angles.

(154) (a) How many degrees are there in 240 minutes? (b) How many seconds? Ans. $\begin{cases} (a)\ 4° \\ (b)\ 14{,}400″ \end{cases}$

(155) Draw an obtuse angle, a right angle, and an acute angle. State the name of each angle by using letters to designate them.

(156) Draw a rhombus and then draw a rectangle having the same area.

(157) Can a quadrilateral be formed with lines whose lengths are 20 inches, 9 inches, 4 inches, and 7 inches? Give reasons?

(158) A sheet of zinc measures 11½ inches by 2½ feet. How many square inches does it contain? Ans. 345 sq. in.

(159) If the zinc in the last example weighs 5¼ pounds, what is its weight per square foot? Ans. 2.19 lb.

(160) How many boards 16 feet long and 5 inches wide would be required to lay a floor measuring 15 × 24 feet?
Ans. 54 boards.

(161) A lot of land is in the shape of a trapezoid. It is 16 rods long, 9 rods wide at the front, and 6 rods wide at the back. The front and back being parallel, what part of an acre does the lot contain? Ans. ¾ of an acre.

(162) The accompanying figure shows the floor plan of

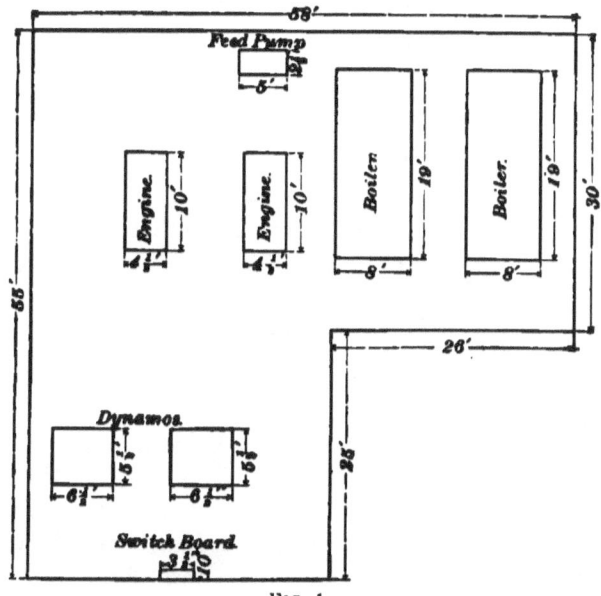

Fig. 1.

an electric-light station. From the dimensions given, calculate the number of square feet of unoccupied floor space.
Ans. 2,059.08 sq. ft.

(163) A sidewalk 10 feet wide extends around a city block 528 feet long and 352 feet wide. Assuming that there is no space between the walk and the buildings, how many square yards does the walk contain? Ans. 2,000 sq. yd.

(164) How many square yards of plastering will be required for the four side walls of a hall 90 feet long, 50 feet wide, and 20 feet high, with four doors $5\frac{1}{2} \times 10$ feet and fourteen windows 5×11 feet? There is a baseboard 9 inches high. Ans. 490.72 sq. yd.

(165) A triangle has three equal angles; what is it called?

(166) If a triangle has two equal angles, what kind of a triangle is it?

(167) Can a triangle be formed with three lines whose lengths are 12 inches, 7 inches, and 4 inches? Give reasons for your opinion.

(168) (*a*) What is the altitude of an equilateral triangle whose sides are each 6 feet? (*b*) What is its area?

Ans. $\begin{cases} (a)\ 5.196\ \text{ft.} \\ (b)\ 15.588\ \text{sq. ft.} \end{cases}$

(169) In a triangle ABC, angle $A = 23°$, and $B = 32°\ 32'$; what does angle C equal? Ans. $C = 124°\ 28'$.

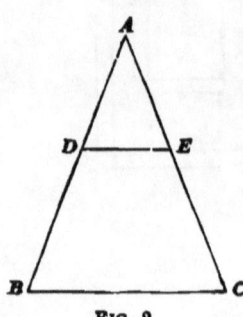

FIG. 2.

(170) In the figure, if $AD = 10$ inches, $AB = 24$ inches, and $BC = 13\frac{1}{2}$ inches, how long is DE, DE being parallel to BC? Ans. $DE = 5.625$ in.

(171) An engine room is 52 feet long and 39 feet wide. How many feet is it from one corner to a diagonally opposite one, measured in a straight line?
Ans. 65 ft.

(172) A ladder 24 feet long rests against a house with its upper end 8 feet from the ground. How far on the ground is the lower end of the ladder from the house? Ans. 22 ft. $7\frac{1}{2}+$ in.

(173) (*a*) Show why it is that the area of a triangle equals one-half the product of the base by the altitude. (*b*) Does it make any difference which side is taken as the base?

(174) (*a*) The area of an isosceles triangle is 200 square inches. If its altitude is 20 inches, how long is its base? (*b*) What is the length of one of its equal sides?

Ans. $\begin{cases} (a)\ 20\ \text{in.} \\ (b)\ 22.36\ \text{in.} \end{cases}$

(175) In an equilateral heptagon one of the sides equals 3 inches; what is the length of the perimeter? Ans. 21 in.

(176) The perimeter of a regular decagon is 40 inches; what is the length of the side? Ans. 4 in.

(177) What is one angle of a regular dodecagon equal to?
Ans. 150°.

(178) The area of a regular pentagon is 43 square inches. If one side is 5 inches long, what is the perpendicular distance from the center to one side? Ans. 3.44 in.

MENSURATION.

(179) It is required to make a miter-box, in which to cut molding to fit around an octagon post. At what angle with the side of the box should the saw run? Ans. $67\frac{1}{2}°$.

(180) Calculate the area of the irregular polygon, Fig. 3. The dimensions are to be obtained by measuring. Ans. $1.78 +$ sq. in.

(181) An angle inscribed in a circle intercepts one-fourth the circumference. How many degrees are there in the angle? Ans. 45°.

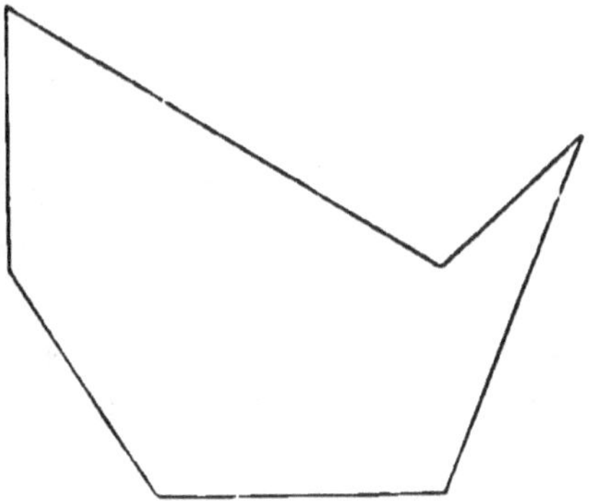

FIG. 3.

(182) If the distance between two opposite corners of a hexagonal nut is two inches, what is the distance between two opposite sides? Ans. $1.732 +$ in.

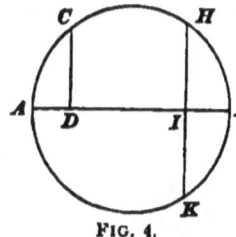

FIG. 4.

(183) In the accompanying figure, if the distance $B I$ is 6 inches and $H K$ 18 inches, what is the diameter of the circle? Ans. 19.5 in.

(184) In the same figure, if the diameter $A B = 32\frac{1}{2}$ feet, and the distance $I B = 8$ feet, what is the length of the chord $H K$? Ans. 28 ft.

(185) The trunk of a tree measures 7.854 feet around it; what is its diameter? Ans. $2\frac{1}{2}$ ft.

(186) How many revolutions will a 72-inch locomotive driver make in going one mile? Ans. 280.112 revolutions.

(187) A pipe has an internal diameter of 6.06 inches; what is the area of a circle having this diameter? Ans. 28.8427 sq. in.

(188) The area of a circle having a diameter corresponding to the internal diameter of a certain pipe is 113.0973 square inches. What is the outside diameter, the pipe being $\frac{3}{4}$ inch thick? Ans. $13\frac{1}{2}$ inches.

MENSURATION.

(189) How long must the arc of a circle be to contain 12°, supposing the radius of the circle to be 6 inches?
Ans. 1.25664 in.

(190) Calculate the area of a flat circular ring whose outside diameter is 22 inches, and whose inside diameter is 21 inches. Ans. 33.7722 sq. in.

(191) Find the area of a segment of a circle whose diameter is 56 inches, the height of the segment being 5 inches.
Ans. 108.5 sq. in.

(192) What are the dimensions of the end of the largest square bar that can be planed from an iron bar 2 inches in diameter? Ans. 1.4142 in. square.

(193) What is the area of the sector of a circle 15 inches in diameter, the angle between the two radii forming the sector being $12\frac{1}{2}°$? Ans. 6.1359 sq. in.

(194) (*a*) What would be the length of the side of a square metal plate having an area of 103.8691 square inches? (*b*) What would be the diameter of a round plate having this area? (*c*) How much shorter is the circumference of the round plate than the perimeter of the square plate?

Ans. $\begin{cases} (a)\ 10.1916\ \text{in.} \\ (b)\ 11\frac{1}{2}\ \text{in.} \\ (c)\ 4.638\ \text{in.} \end{cases}$

(195) A plate 46 inches long is to be rolled into a shell; what will be the diameter of the shell if 5 inches are allowed for lap? Ans. 13.05 in.

(196) What is the convex area in square feet of a section of a smokestack 26 inches in diameter and $10\frac{1}{4}$ feet long?
Ans. 71.4714 sq. ft.

(197) Find the area in square feet of the entire surface of a hexagonal column 12 feet long, each edge of the ends of the column being 4 inches long? Ans. 24.5774 sq. ft.

(198) Find the cubical contents of the above column in cubic inches. Ans. 5,985.9648 cu. in.

MENSURATION.

(199) Compute the weight per foot of an iron boiler tube 4 inches outside diameter and 3.73 inches inside diameter, the weight of the iron being taken at 28 pound per cubic inch. Ans. 5¼ lb.

(200) The dimensions of a return-tubular boiler are as follows: Diameter, 60 inches; length between heads, 16 feet; outside diameter of tubes, 3½ inches; number of tubes, 64; distance of mean water-line from top of boiler, 18 inches. (*a*) Compute the steam space of the boiler in cubic feet. (*b*) Determine the number of gallons of water that will be required to fill the boiler up to the mean water level?

Ans. $\begin{cases} (a) \ 79.2 \text{ cu. ft.} \\ (b) \ 1,246 \text{ gal., nearly.} \end{cases}$

(201) The cylinders of a compound engine are 19 and 31 inches in diameter, and the stroke is 24 inches; if the clearance at *each* end of the small cylinder is 14% of the *stroke*, and in the large cylinder 8% of the *stroke*, (*a*) what is the total volume in cubic feet of the steam in the small cylinder during one stroke? (*b*) In the large cylinder? (*c*) What is the ratio between the two?

Ans. $\begin{cases} (a) \ 4.489 \text{ cu. ft.} \\ (b) \ 11.321 \text{ cu. ft.} \\ (c) \ \text{Ratio} = 2.522 : 1. \end{cases}$

(202) Find the volume of a triangular pyramid 10 inches high; the base is an equilateral triangle, and one edge measures 10 inches. Ans. 144.336 cu. in.

(203) The slant height of a square pyramid is 25 inches, and one edge of its base is 16 inches. Find its altitude by the principle of the right-angled triangle. Ans. 23.6854 in.

(204) The length of the circumference of the base of a cone is 18.8496 inches, and its slant height is 10 inches. Find the area of the entire surface of the cone.
Ans. 122.5224 sq. in.

(205) If the altitude of the above cone were 9 inches, what would be its volume? Ans. 84.8232 cu. in.

(206) A square vat is 11 feet deep, 15 feet square at the top, and 12 feet square at the bottom. How many gallons will it hold? Ans. 15,058.29 gal.

(207) How many pails of water would be required to fill the vat, the pail having the following dimensions: Depth, 11 inches; diameter at the top, 12 inches; diameter at the bottom, 9 inches? Ans. 3,627.28.

(208) Required the area of the convex surface of the frustum of a hexagonal pyramid whose slant height is 32 inches, the perimeter of the lower base measuring 48 inches, and of the upper base 36 inches. Ans. 1,344 sq. in.

(209) It is desired to find the number of gallons that a certain tank will hold. By measuring with a tape measure, it is found to be 190 inches in circumference at the bottom and $170\frac{1}{2}$ inches in circumference at the top. The depth of the tank is 7 feet, and the thickness of the sides $1\frac{1}{4}$ inches. Make the calculation in round numbers. Ans. 861 gal.

NOTE.—To obtain the area of the upper and lower bases, first find the outside diameters and then deduct the thickness of the walls, or $2 \times 1\frac{1}{4}$ inches.

(210) Find (*a*) the area of the surface, and (*b*) the cubical contents of a ball $22\frac{1}{2}$ inches in diameter.

Ans. $\begin{cases} (a)\ 1{,}590.435\ \text{sq. in.} \\ (b)\ 5{,}964.1313\ \text{cu. in.} \end{cases}$

(211) What is the volume of a ball whose surface has an area of 201.0624 square inches? Ans. 268.0832 cu. in.

(212) (*a*) What is the volume and area of a cylindrical ring whose outside diameter is 16 inches and inside diameter 13 inches? (*b*) If made of cast iron, what is its weight? Take the weight of 1 cubic inch of cast iron as .261 pound.
Ans. Weight = 21 lb.

(213) The volume of a certain cylindrical ring is 144.349 cu. in., its length before bending (see line D in Fig. 61, Art. **432**) was 20.42 inches; what is the area of its surface?
Ans. 192.45 sq. in.

MECHANICS.
(SEE ARTS. 433–558.)

(214) (*a*) What is a molecule ? (*b*) An atom ?

(215) If a body has an average velocity of 40 feet per second, how far will it travel in 14 minutes ? Ans. $6\tfrac{4}{11}$ miles.

(216) Show how to represent a force by a line.

(217) In Fig. 78, Art. **491,** let the distance $F c$ be 21″, and $F b$ 3½″; what weight will a force of 85 lb. applied at P raise ? Ans. 510 lb.

(218) What must be the speed of the driver pulley in order that the driven may make 80 R. P. M. and be 28″ in diameter, the diameter of the driver being 21″ ?
Ans. $106\tfrac{2}{3}$ R. P. M.

(219) The number of teeth in a spur gear is 50 and the pitch is 1½″; (*a*) what is the pitch diameter ? (*b*) What is the outside diameter? Ans. $\begin{cases} (a)\ 23.87''. \\ (b)\ 24.77''. \end{cases}$

(220) The driving gear has 45 teeth and the driven 180; if the driver makes 212 R. P. M., how many will the driven make ? Ans. 53 R. P. M.

(221) What pressure can be exerted by a force of 24 lb. on a half-inch screw which has 13 threads per inch, the distance from the center of the screw to the point on the handle where the force is applied being 11″ ?
Ans. 21,563.94 lb.

(222) A ball weighing 5 lb. revolves in a circle whose radius is 32″ at the rate of 350 R. P. M.; what is the pull on the support caused by the ball ? Ans. $555\tfrac{1}{2}$ lb.

(223) A body weighing 2 lb. has a velocity of 600 ft. per sec.; what is its kinetic energy ? Ans. 11,194 ft.-lb.

(224) What should be the width of a double leather belt to transmit 150 horsepower, when the belt has a velocity of 3,000 feet per minute, and has 7 feet of its length in contact with the smaller pulley, whose diameter is 63"? Give width to nearest $\frac{1}{4}$". Ans. 29.5".

(225) (*a*) What are the three states of matter? (*b*) Name some of the general properties of matter; (*c*) some of the specific properties.

(226) If a man could run a mile at the average rate of 100 yards in 12 seconds, how long would it take him?
Ans. 3 min. 31.2 sec.

(227) What is meant by "center of gravity"?

(228) (*a*) Why is crowning usually given to the face of a pulley? (*b*) Why should high-speed pulleys be balanced?

(229) At what speed must the engine run when the diameter of the band-wheel is 13 feet and of the main pulley 91", if the speed of the main shaft is to be 108 R. P. M.?
Ans. 63 R. P. M.

(230) The pitch of a gear is $2\frac{1}{2}$", and the number of teeth is 192; what is the pitch diameter? Ans. 152.79".

(231) How many revolutions per minute must the driving gear make if it has 18 teeth, and the driven has 81 teeth and makes 80 R. P. M.? Ans. 360 R. P. M.

(232) The nuts on a cylinder head are tightened by means of a wrench 26" long. The threads in the nuts are 8 to the inch, and the efficiency of the screw is 40%. What pressure will the nut exert against the head when a force of 60 lb. is applied to the end of the wrench? Ans. 31,365.7 lb.

(233) What do you understand by specific gravity?

(234) If the center of gravity of a section of an engine fly-wheel rim is 6 ft. $1\frac{3}{4}$ in. from the center of the shaft, and the weight of the rim is 13,000 pounds, what is its kinetic energy when making 150 R. P. M.? Ans. 1,883,661.7 ft.-lb.

(235) What should be the width of a single leather belt to transmit $2\frac{1}{2}$ horsepower when the belt has a velocity of

2,000 feet per minute? The diameter of the smaller pulley is 14″, and the belt has 18″ of its length in contact with it.

Ans. 1⅛″.

(236) (*a*) What is meant by inertia? (*b*) By weight? (*c*) How is weight measured?

(237) The speed of a certain belt is 3,000 ft. per min.; if it drives a 48″ pulley, how long will it take the pulley to make 100 revolutions?   Ans. 25.13 sec., nearly.

(238) Find the point of suspension of a rectangular cast-iron lever 4 ft. 6 in. long, 2 in. deep, and ¾ in. thick, having weights 47 and 71 pounds hung from each end, in order that there may be equilibrium. Take the weight of a cubic inch of cast iron as .261 lb.   Ans. $\begin{cases} \text{Short arm} = 22.343''. \\ \text{Long arm} = 31.657''. \end{cases}$

(239) When two pulleys are used to transmit power, which is called the driven and which the driver?

(240) The driver is 2 feet in diameter and the driven 32″; if the driven makes 63 R. P. M., how many must the driver make?   Ans. 84 R. P. M.

(241) A certain gear has a pitch of 1¼″, and its pitch diameter is 11.48″; how many teeth are in the gear?

Ans. 32 teeth.

(242) A fly-ball governor is designed to run at 88 R.P.M. The speed of the engine is 200 R. P. M. The diameter of the governor pulley is 8″; the number of teeth in the bevel gear which it turns, 44, and the number of teeth in the other bevel gear, 75; what must be the diameter of the pulley on the crank-shaft which drives the governor belt?   Ans. 6″.

(243) A bookbinder has a press, the screw of which has 4 threads to the inch. It is worked by a lever 15″ long, to which is applied a force of 25 lb.; (*a*) what will be the pressure if the loss by friction is 5,000 lb.? (*b*) What would be the theoretical pressure?   Ans. $\begin{cases} (a)\ 4,424.8\ \text{lb.} \\ (b)\ 9,424.8\ \text{lb.} \end{cases}$

(244) A cubic foot of a certain kind of wood weighs 51 lb.; what is its specific gravity?   Ans. .816.

(245) The piston of an engine weighs 325 lb., including the piston rod; what is its kinetic energy when moving at the rate of 660 ft. per min.?   Ans. 611.4 ft.-lb., nearly.

(246) What horsepower can be safely transmitted by a gear whose pitch is 1″, a point on the pitch circle having a velocity of 1,200 ft. per min.?   Ans. 12 H. P.

(247) (a) What is motion? (b) Velocity? (c) Rest? (d) Can a body be in motion with respect to one object and at rest with respect to another? Explain fully.

(248) (a) What is force? (b) Name several kinds of forces.

(249) Find by measurement the center of gravity of a triangle whose sides are 4″, 5″, and 6″ long.
   Ans. $1\frac{3}{32}″$ from 6″ side.

(250) The driving pulley makes 40 R. P. M.; the driven makes 60 R. P. M., and is 36″ in diameter. What is the diameter of the driver?   Ans. 54″.

(251) The diameter of the band-wheel of an engine is 12 ft.; of the main pulley, 8 ft.; of a driving pulley on the main shaft, 20″; of the driven pulley on the countershaft, 6″; of the driver on the countershaft, 6″, and of the driven on an emery-wheel spindle, 4″; if the engine makes 80 R. P. M., (a) what is the speed of the main shaft? (b) Of the countershaft? (c) Of the emery wheel?

Ans. $\begin{cases} (a)\ 120\ \text{R. P. M.} \\ (b)\ 400\ \text{R. P. M.} \\ (c)\ 600\ \text{R. P. M.} \end{cases}$

(252) The pitch diameter of a gear is 34.15″; if the pitch is $1\frac{3}{8}″$, how many teeth has the gear?   Ans. 78 teeth.

(253) In order to raise a weight, a combination of a fixed and movable pulley is used; if a force of 225 lb. be applied to the free end of the rope, what load will it raise?
   Ans. 450 lb.

(254) If the power moves through a distance of 5 ft. 6 in. while the weight is moving 6 in., (a) what is the velocity

ratio of the machine? (*b*) What weight would a force of 5 lb. applied to the power arm raise?   Ans. $\begin{cases} (a)\ 11. \\ (b)\ 55\ \text{lb.} \end{cases}$

(255) In the last example, if the efficiency were 65%, what weight could the machine raise?   Ans. 35.75 lb.

(256) How many cubic inches of platinum will it take to weigh 10 lb.?   Ans. 12.86 cu. in., nearly.

(257) (*a*) For what are belts used? (*b*) What is a single belt? (*c*) A double belt?

(258) What horsepower can be safely transmitted by a gear whose pitch is 1.57", pitch diameter 30", and which makes 100 revolutions per minute?   Ans. 19.36 H. P.

(259) (*a*) What is uniform motion? (*b*) What is variable motion? (*c*) If a body moves 10 feet the first second, 12 feet the second second, 15 feet the third second, etc., is its motion uniform or variable, and why?

(260) What three conditions are required to be known in order to compare forces?

(261) A steel rod $\frac{3}{4}$" in diameter has on one end a cast-iron spherical ball 5" in diameter; if the length of the rod is 40" between the ball and the end, where is the center of gravity?

SUGGESTION.—First calculate the weights of the ball and rod by the aid of the specific gravity tables.
   Ans. 6.427" from the center of the ball.

(262) The driving pulley makes 240 R. P. M., and the driven pulley 180 R. P. M.; if the diameter of the driven pulley is 30", what is the diameter of the driver?   Ans. 22$\frac{1}{2}$".

(263) In a train of gears used to raise a weight of 6,000 lb. in a manner similar to that shown in Fig. 83, Art. **500**, the diameters of the drivers and belt pulley are 18", 12", 15", and 12", and of the pinions and drum 6", 5", 8", and 3"; what force must be applied to the belt to raise the weight, if 20% of the total force is lost through friction?
   Ans. 138$\frac{8}{9}$ lb.

(264) The pitch diameter of a gear is 24.16", and the number of teeth is 38; what is the pitch?   Ans. 1.997".

MECHANICS.

(265) It is required to raise a load of 1,890 lb. by means of a block and tackle which has four fixed and four movable pulleys; what force is required to be applied to the free end of the rope? Ans. $236\frac{1}{4}$ lb.

(266) In a block and tackle, the theoretical power necessary to raise a weight of 1,000 lb. is 50 lb.; (a) what is the velocity ratio? (b) If the actual power necessary to raise the load is 95 lb., what is the efficiency? Ans. $\begin{cases} (a)\ 20. \\ (b)\ 52.63\%. \end{cases}$

(267) A piece of lead is $\frac{1}{2}''$ in diameter and $10''$ long; how much does it weigh? Ans. 12.91 oz.

(268) If the distance between the centers of the crank-shaft and main shaft is 38 ft., and the diameters of the band-wheel and main pulley are 11 ft. and 7 ft., respectively, what must be the length of the main belt? Ans. 105 ft. $3''$.

(269) The entire solar system is moving through space at the rate of 18 miles per second; (a) what is its velocity in miles per hour? (b) How far will it go in one day?
Ans. $\begin{cases} (a)\ 64{,}800\ \text{mi. per hr.} \\ (b)\ 1{,}555{,}200\ \text{mi.} \end{cases}$

(270) (a) What is a path of a body? (b) What line do we measure when we wish to find the distance a body has traveled.

(271) (a) How are forces measured? (b) What kind of an effect do forces always tend to produce?

(272) It is required to raise a weight of 1,500 lb. by means of a lever like that shown in Fig. 71, Art. **484**. The length of the lever is 4 ft., and the distance from the fulcrum to the weight is $4''$; what force will it be necessary to apply? Ans. $136\frac{4}{11}$ lb.

(273) Had the lever in the above example been like that shown in Fig. 72, Art. **484**, what force would have been required? Ans. 125 lb.

(274) The band-wheel of an engine is 10 feet in diameter; what must be the diameter of the pulley on the main shaft in order to make 110 R. P. M., if the band-wheel makes 88 R. P. M.? Ans. 8 ft.

MECHANICS.

(275) (*a*) What is a spur gear? (*b*) A miter gear? (*c*) A bevel gear?

(276) The pitch diameter of a gear is 36.56″, and the number of teeth is 42; what is the pitch?   Ans. 2.735″.

(277) The length of an inclined plane is 400 feet, and the height is 45 feet; what force acting parallel to the plane will be required to pull up the plane a weight of 4,000 lb.?

Ans. 450 lb.

(278) (*a*) What is meant by the velocity ratio of a machine? (*b*) By the efficiency?

(279) If a coil of brass wire weighs 10 lb., and the diameter of the wire is $\frac{1}{10}''$, how long is the wire?

Ans. 896 ft., nearly.

(280) The diameters of two pulleys are 14″ and 18″, and the distance between their centers is 14 feet; what must be the length of a belt to drive these pulleys?   Ans. 32 ft. 4″.

(281) A velocity of 30 miles per hour corresponds to how many feet per second?   Ans. 44 ft. per sec.

(282) The stroke of a steam engine is 28″, and it makes 1,500 strokes in 6 minutes; what is the velocity of the piston in feet per second?   Ans. $9\frac{13}{4}$ ft. per sec.

(283) State the three relations between force and motion.

(284) A pulley on the main shaft is 40″ in diameter, and makes 120 R. P. M.; what must be the diameter of a pulley on the countershaft that is to make 160 R. P. M.?

Ans. 30″.

(285) (*a*) What is a rack? (*b*) A worm-wheel? (*c*) A worm?

(286) (*a*) What distinguishes the epicycloidal teeth from the involute teeth? (*b*) Name two advantages which the latter possess over the former.

(287) An inclined plane has a length of 1,200 feet and a height of 125 feet. It is required to pull a load of 50,000 lb. up this plane. A block and tackle having 6 fixed and 6 movable pulleys is stationed at the top of the plane, and the weight end of the rope is attached to the load. If the rope which connects the block to the load is parallel to the plane,

what force will it be necessary to exert on the free end of the rope to pull up the load, no allowance being made for friction?
Ans. 434 lb.

(288) (*a*) What do you understand by centrifugal force? (*b*) By centripetal force?

(289) Define (*a*) work; (*b*) horsepower; (*c*) kinetic energy; (*d*) potential energy.

(290) Two pulleys have diameters of 8″ and 20″; the distance between their centers being 19 ft. 3″, what must be the length of a belt to drive them? Ans. 42 ft. $3\frac{1}{2}$″.

(291) Two bodies starting from the same point, move in opposite directions, one at the rate of 11 feet per second, and the other, 15 miles per hour; (*a*) what will be the distance between them at the end of 8 minutes; (*b*) How long before they will be 825 feet apart? Ans. $\begin{cases} (a)\ 3\ \text{miles.} \\ (b)\ 25\ \text{seconds.} \end{cases}$

(292) A railroad train runs 2 miles in 2 minutes and 10 seconds; what is its average velocity in feet per second?
Ans. 81.23 ft. per sec.

(293) Why is it difficult to jump from a rowboat into the water?

(294) A compound lever, similar to the one shown in Fig. 77, Art. **490,** is required to lift a weight of 1,250 lb. The lengths of the power arms PF are 30″, 20″, 10″, and 15″, of the weight arms WF 6″, 5″, 4″, and 7″; what force will be required? Ans. $11\frac{2}{3}$ lb.

(295) The driving pulley is 20″ in diameter and driven 16″; if the driver makes 150 R. P. M., how many will the driven make? Ans. $187\frac{1}{2}$ R. P. M.

(296) How is the diameter of a gear measured?

(297) The driving gear makes 100 and the driven, 40 R. P. M.; if the driven has 60 teeth, how many has the driver? Ans. 24.

(298) The base of an inclined plane is 80 feet long, and the height of the plane is 50 feet; what force exerted parallel to the base will raise a load of 750 lb.? Ans. $468\frac{3}{4}$ lb.

(299) What is the centrifugal force of the counterweight of a steam engine, the counterweight weighing 128 lb., and its center of gravity being 8¾" from the center of the shaft? The crank makes 180 R. P. M. Ans. 1,028.16 lb.

(300) How much work can be done by 20 cubic feet of water falling from a height of 50 feet? Ans. 62,500 ft.-lb.

(301) What horsepower can be transmitted by a single leather belt 5" wide, which runs at the rate of 1,960 ft. per min.? The diameter of the smaller pulley is 15" and the length of the arc of contact is 21"? Ans. 11.4 H. P., nearly.

(302) It is required to raise a weight of 18,000 lb. by means of a screw having 3 threads per inch; if the length of the handle is 15", and there is a loss of 10,000 lb. due to friction, etc., what force will it be necessary to apply to the handle?
Ans. 99 lb., nearly.

(303) The fly-wheel of an engine is 9 feet in diameter (outside); if the fly-wheel makes 100 R. P. M., how many miles will a point on the rim travel in 1¼ hours?
Ans. 40.16¼ miles.

(304) Suppose that an air gun can throw a ball with a velocity of 100 feet per second, and that a man standing on a railroad train, which is moving at the rate of 100 feet per second, were to fire the gun in a direction exactly opposite to that in which the train is moving, what would become of the ball? Why?

(305) If the distance between the center line of the handle and the axis of the drum shown in Fig. 79, Art. **491**, is 14½", and the diameter of the drum is 5", what load will a force of 30 lb. exerted on the handle raise? Ans. 174 lb.

(306) A pulley on the main shaft is 42" in diameter, and makes 108 R. P. M.; what will be the speed of the countershaft if the driven pulley is 36" in diameter?
Ans. 126 R. P. M.

(307) What is (*a*) the pitch circle? (*b*) The pitch of a gear?

(308) The driving gear makes 360 R. P. M., and the driven 170 R. P. M.; if the driver has 34 teeth, how many has the driven? Ans. 72 teeth.

(309) Name some particular use in the engine room or shop to which you have seen the inclined plane put.

(310) The mean diameter of the rim of an engine cast-iron fly-wheel is 9 ft. $10\frac{3}{4}''$, its width is $22''$, and its thickness $2\frac{1}{2}''$; what is the centrifugal force when running at 210 R. P. M.? Ans. 397,450 lb.

(311) Assuming the average pressure upon the piston of a steam engine to be 41.38 pounds per square inch, what is the horsepower? The diameter of the piston is $10''$; the stroke $16''$, and the number of strokes per minute 450.

Ans. 59.091 H. P., nearly.

(312) What horsepower can be transmitted by a $20''$ double leather belt which has a velocity of 2,800 ft. per min.? The diameter of the smaller pulley is 4 ft., and the belt has 5 ft. 9 in. of its length in contact with it. Ans. 99.4 H. P.

MECHANICS.

(SEE ARTS. 559-655.)

(313) State Pascal's law.

(314) A cylinder fitted with a piston is used as a lifting cylinder by passing a rope over a pulley and fastening one end to the piston rod. The piston is moved by means of water obtained from the city reservoir, and a gauge attached to a pipe near the cylinder shows the pressure to be 90 lb. per sq. in. The diameter of the cylinder is 19 in. and of the pipe ½ in. If friction be neglected, how great a weight can be raised? Ans. 25,517.6 lb.

(315) Give Mariotte's law.

(316) What do you understand by (*a*) the tensile strength of a material? (*b*) The working stress?

(317) A close-link wrought-iron chain is made from $\frac{5}{8}''$ iron; what is the greatest safe load that it will carry?
Ans. 1,687.5 lb.

(318) What is the allowable working load for a steel-wire rope $5\frac{1}{4}''$ in circumference? Ans. 27,562.5 lb.

(319) What steady force is required to shear a steel crank-pin which is 6" in diameter? Ans. 1,696,464 lb.

(320) What must be the diameter of a wrought-iron shaft to transmit 40 horsepower when running at 116 revolutions per minute, no power being taken off between bearings. Ans. 2.89". A shaft $2\frac{15}{16}''$ in diameter would be used, that is, 3" round iron turned down.

(321) In Fig. 104, Art. **562**, suppose that the area of the piston *c* is .6 sq. in.; of *d*, 3 sq. in.; of *e*, 6 sq. in.; of *f*, 2 sq. in.; of *a*, 14 sq. in., and of *b*, 9 sq. in.; if a force of

24 lb. be applied to *c*, what must be the forces applied to the other pistons to counterbalance the force at *c*, neglecting the weight of the water and the weight of the pistons?

Ans. $\begin{cases} \text{At } d, & 120 \text{ lb.} \\ \text{At } e, & 240 \text{ lb.} \\ \text{At } f, & 80 \text{ lb.} \\ \text{At } a, & 560 \text{ lb.} \\ \text{At } b, & 360 \text{ lb.} \end{cases}$

(322) The upper base of a cylinder submerged in water is 40 ft. below the surface. The diameter of the cylinder is 20 inches, the altitude 36 inches, and the bases are parallel. If the bases are horizontal, what is (*a*) the upward pressure of the water on the cylinder? (*b*) The downward pressure?

Ans. $\begin{cases} (a) & 5,863.39 \text{ lb.} \\ (b) & 5,454.32 \text{ lb.} \end{cases}$

(323) (*a*) What is a barometer? (*b*) What is the essential difference between a mercurial and an aneroid barometer?

(324) The smallest section of a connecting-rod is 3.5 sq. in.; what is the unit stress when subjected to a tensile stress of 12,400 lb.? Ans. 3,543 lb. per sq. in., nearly.

(325) What load may be safely carried by a hemp rope 4″ in circumference? Ans. 1,600 lb.

(326) What load can be safely sustained by a round wooden pillar, 8″ in diameter and 10 ft. long, having both ends flat? Ans. 13⅓ tons.

(327) What force is required to punch a 1″ hole through a wrought-iron plate $\frac{7}{16}$″ thick? Ans. 54,978 lb.

(328) What horsepower will a 1¼″ wrought-iron shaft transmit when running at 180 revolutions per minute, an average amount of power being taken off between the bearings? Ans. 12.49 H. P.

(329) Does the shape of the vessel have any effect in regard to the pressure exerted by a liquid upon its bottom?

(330) In Fig. 111, Art. **576**, suppose that the diameter of the piston *a* is 2″, and of *b* 7½″; if a weight of 400 lb. be

MECHANICS.

laid upon b, what weight must be applied to a to balance the weight on b ? Ans. 28.44 + lb.

(331) A vessel contains 42 cu. ft. of coal gas having a tension of $20\frac{1}{4}$ lb. per sq. in.; what will be the new tension when allowed to communicate with a perfectly empty vessel whose volume is 14 cu. ft.? Ans. 15.19 lb. per sq. in., nearly.

(332) What safe steady load can be sustained by a $1\frac{1}{4}''$ round wrought-iron bar, the load producing a tensile stress? Ans. 21,205.2 lb.

(333) What load can a hemp rope 6" in circumference carry with safety? Ans. 3,600 lb.

(334) What load will a hollow cast-iron pillar support with safety if the pillar is 20 ft. long, outside diameter 14", inside diameter $11\frac{1}{2}''$, and both ends are fixed?
Ans. 219.24 tons.

(335) What force is required to punch a hole $1\frac{1}{4}''$ in diameter through a $\frac{3}{4}''$ steel plate? Ans. 212,058 lb.

(336) A vessel is filled with water to the depth of 18"; if the area of the bottom is 46 sq. in., what is the total pressure on the bottom? Ans. 29.95 lb.

(337) Why does water seek its level?

(338) The volume of steam in an engine cylinder is $\frac{5}{8}$ cu. ft. at cut-off, and its pressure is 94.7 lb. per sq. in.; what will be its pressure when the volume has increased to $2\frac{1}{2}$ cu. ft.? Ans. 23.675 lb. per sq. in.

(339) A bar of steel having a cross-section of $1\frac{3}{4}'' \times 3''$ is subjected to a tensile stress; if the stress is suddenly applied, what is the greatest load that it will safely carry?
Ans. 31,500 lb.

(340) A load of 2,400 lb. is to be raised by means of a hemp rope; what should be the circumference of the rope? Ans. 4.9'.

(341) Regarding the piston rod of a steam engine as a pillar which has one end flat and the other round, what should be the greatest diameter of the piston if the rod is 4 ft. 8 in. long, $3\frac{1}{2}$ in. in diameter, and the greatest steam

pressure is not to exceed 100 lb. per sq. in.? The rod is made of wrought iron. Ans. 25″, nearly.

(342) (a) What is cold-rolled shafting? (b) Bright shafting? (c) Black shafting?

(343) A tank contains cylinder oil having a specific gravity of say .92. This tank is connected to a four-gallon can by means of a pipe whose internal diameter is ⅜″. If the vertical distance between the level of the oil in the tank and the end of the pipe is 8 ft. 7 in., what is the pressure per square inch at the end of the pipe?
Ans. 3.427 lb. per sq. in.

(344) In Fig. 113 (see examples following Art. **578**), suppose that the diameter of the plunger A is 12″, and of the plunger C of the pump B ⅜″; if a force of 160 lb. be applied to C by means of the lever D, how great a weight can the plunger A raise if the weight of A is 600 lb.?
Ans. 58,382.4 lb.

(345) The volume of steam in an engine cylinder at the beginning of compression is 1.11 cu. ft., and its pressure is 18 lb. per sq. in. At the end of compression the volume is .3 cu. ft.; what is the final pressure?
Ans. 66.6 lb. per sq. in.

(346) What should be the least area of one of the 14 wrought-iron cylinder head stud bolts if the diameter of the cylinder is 19″, and the greatest steam pressure is 180 lb. per sq. in.? Assume that the studs are subjected to shocks.
Ans. .729 sq. in.

(347) What should be the circumference of a hemp rope to safely sustain a load of 4,200 lb.? Ans. 6¼″.

(348) Regarding the connecting-rod of a steam engine as a pillar with two round ends, what is the greatest force that may be exerted on the cross-head if the connecting-rod is made of wrought iron, is 10 ft. long. and has a rectangular cross-section 6″ by 2½″? Ans. 35,489 lb.

(349) If you were to order some 2½″ bright-turned shafting, what size would you expect to get?

MECHANICS.

(350) A tank having the shape of a frustum of a cone is filled with water. If the diameter of the large end is 8 ft., of the small end 6 ft., and the perpendicular distance between the two ends is 10", (*a*) what is the pressure on the bottom when the large end is down? (*b*) When the small end is down?

Ans. { (*a*) 2,618 lb.
{ (*b*) 1,473 lb., nearly.

(351) What is the total pressure on all of the six sides of a cube which has been sunk in the water until its top is 50 ft. below the surface? One edge of the cube measures 2 ft., and its upper base is parallel with the water level.

Ans. 76,502 lb.

(352) A vessel contains 25 cu. ft. of air having a pressure of 45 lb. per sq. in. When allowed to communicate with a second vessel which is entirely empty, the pressure falls to 15 lb. per sq. in. What is the volume of the second vessel? Ans. 50 cu. ft.

(353) A steam cylinder is 44" in diameter and sustains a steam pressure of 100 lb. per sq. in. The diameter of the cylinder head studs is $1\frac{3}{8}$" and the area at the bottom of the thread is 1.057 sq. in. How many wrought-iron studs are required? Assume that the studs are subjected to shocks. Ans. 29 studs.

(354) An iron-wire rope 4" in circumference is used on a crane for hoisting loads; what is the greatest load that the rope will sustain with safety? Ans. 9,600 lb.

(355) A cast-iron rectangular cantilever beam having a cross-section of $1\frac{1}{2}$" wide by $2\frac{1}{2}$" deep is 4 ft. 8" long; how great a weight will the beam sustain at its end?

Ans. 201 lb., nearly.

(356) What horsepower will a $2\frac{7}{16}$" steel shaft transmit, when running at 120 revolutions per minute, pulleys being carried between the bearings to distribute the power along the line? Ans. 20.445 H. P.

(357) A board 8" × 20" and $1\frac{1}{2}$" thick is so placed in water that its flat sides are horizontal; if the distance from the

level of the water to the top of the board is 5 feet, what is the total upward pressure on the board ? Ans. 355.91 + lb.

(358) Will any solid body whose specific gravity is greater than that of water, sink in it to any depth ? Why ?

(359) The volume of steam in an engine cylinder at cut-off is 1.6 cu. ft., and its pressure is 90 lb. per sq. in.; what must be its volume at release when the pressure has fallen to 21 lb.? Ans. 6$\frac{4}{7}$ cu. ft

(360) What should be the least diameter of a wrought-iron bolt that is to resist a sudden pull of 12,000 lb.?
Ans. 1.74″ +.

(361) A steel-wire rope is 4$\frac{3}{4}$″ in circumference; what load will it safely sustain ? Ans. 22,562.5 lb.

(362) A white-pine beam supported at both ends has a rectangular cross-section 8″ wide by 10″ deep; if the beam is 28 ft. long, what total uniform load will it support in safety ? Ans. 6,857¼ lb.

(363) What horsepower can a 10″ wrought-iron crank-shaft transmit, when running at 200 revolutions per minute ?
Ans. 2,857¼.

(364) A vertical cylinder having a diameter of 20″ and a length inside of 36″ is filled with water ; a pipe having a diameter of $\frac{3}{8}$″ is screwed into the upper head and fitted with a piston weighing 10 oz., on which is laid a weight of 25 lb.; if the end of the pipe is 10 ft. above the level of the water in the cylinder, (*a*) what is the pressure per square inch on the top of the cylinder ? (*b*) On the bottom ? (*c*) What equivalent weight laid on the lower cylinder head would replace the pressure it sustains ?
Ans. { (*a*) 236.45 lb. per sq. in.
(*b*) 237.75 lb. per sq. in.
(*c*) 74,691.54 lb.

(365) If, in the last example, a hole one inch in diameter is drilled through the cylinder wall midway of its length, and covered by a flat plate in such a manner that the water can not leak out, what would be the pressure against the plate ? Ans. 186.22 lb.

MECHANICS. 315

(366) The vacuum gauge of a condensing engine indicates 20" of vacuum; what is the pressure in the condenser?
Ans. 4.9 lb. per sq. in.

(367) The volume of the receiver of an air compressor is 300 cu. ft., and the pressure of the air which is contained in it is 52 lb. per sq. in.; if 120 cu. ft. of air, having a pressure of 52 lb. per sq. in., are removed from the receiver, what will be the pressure of the air which remains?
Ans. 31.2 lb. per sq. in.

(368) What is the greatest safe load that may be applied to a stud-link wrought-iron chain if the diameter of the iron from which the link is made is $\frac{1}{2}$"? Ans. 4,500 lb.

(369) It is desired to handle loads up to 14,000 lb. by means of an iron-wire rope; what should be its circumference? Ans. 4.83", nearly.

(370) What is the greatest load that a bar of wrought iron 2" in diameter and 6 ft. long can safely sustain in the middle? The bar is merely supported at its ends.
Ans. 480 lb.

(371) What must be the diameter of a cast-iron crank-shaft to transmit 1.000 horsepower at 80 revolutions per minute? Ans. 10.4 in.

(372) (*a*) What must be the height of the mercury column to indicate a vacuum of 12"? (*b*) Of 18"?

(373) (*a*) What is a stress? (*b*) A strain? (*c*) A unit stress?

(374) The links in a stud-link wrought-iron chain are made from iron $1\frac{3}{8}$" in diameter; what is the greatest safe load that the chain can handle? Ans. 11,883 lb.

(375) A steel-wire rope is used to haul loads up an inclined plane; the greatest stress in the rope is 8,000 lb.; what should be its circumference? Ans. 2.83".

(376) What uniform load can be safely sustained by a steel beam 20 ft. long, 2" wide, and 6" deep? Ans. 4,608 lb.

(377) A 4" steel shaft is to transmit 80 horsepower; how many revolutions per minute must it make if used for transmission only (that is, no power being taken off at intermediate points)? Ans. $81\frac{1}{4}$ R. P. M.

(378) Define tension as applied to gases.

(379) (*a*) What is elasticity? (*b*) Elastic limit? (*c*) What is meant by set?

(380) What safe load may be carried by a close-link wrought-iron chain whose links are made from $\frac{5}{8}''$ iron?
Ans. 4,687.5 lb.

(381) What is the allowable working load for an iron-wire rope 6″ in circumference? Ans. 21,600 lb.

(382) What force is required to shear a wrought-iron strip 4 ft. long and $\frac{1}{2}''$ thick? Ans. 960,000 lb.

(383) A 7″ wrought-iron crank-shaft is to transmit 200 horsepower; how many revolutions per minute must it make?
Ans. 40.8 rev., nearly.

INDEX.

A.

	PAGE
Abstract number	1
Acute angle	120
Addition	4
" of decimals	40
" " denominate numbers	68
" " fractions	29
" Proof of	8
" Rule for	8
" Sign of	4
" table	5
Adjacent angle	120
Aeriform bodies	150
Aggregation, Symbols of	53
Air chamber	244
" compressors	235
" pump	231
Altitude of cone	142
" " frustum of cone or cylinder	144
" " parallelogram	123
" " pyramid	142
" " triangle	129
Amount (in percentage)	56
Aneroid barometer	225
Angle	120
" Acute	120
" Adjacent	120
" Inscribed	133
" Measured by arc	121
" Obtuse	120
" Right	120
" Vertex of	120
Angles or arcs, Measures of	64
Antecedent (of a ratio)	97
Arabic notation	2
Arc of circle	133
" " " To find length of	136
Archimedes, Principle of	220
Area of a surface	124
" " circle	136
" " circular ring	136
" " cone	142
" " cylinder	139
" " cylindrical ring	146
" " frustum of cone	144

	PAGE
Area of frustum of pyramid	144
" " irregular polygon	132
" " parallelogram	124
" " prism	139
" " pyramid	142
" " regular polygon	131
" " sector of circle	137
" " segment of circle	137
" " trapezoid	124
" " triangle	129
Atmosphere, Pressure of	223
Atoms	149
Avoirdupois weight	63

B.

	PAGE
Balanced pulleys	171
Barometer	224
" Aneroid	225
" Mercurial	225
Base (in percentage)	56
" of parallelogram	123
" " plane figure	123
Beams, Transverse strength of	262
Bearings for line shafting, Distance apart of	269
Belts	201
" Double	201
" Horsepower of	202
" Lacing	203
" Length of	201
" Rules for	202
" Single	201
" Width of	202
Bevel gears	175
Black shafting	268
Block (pulley)	183
Bodies, Aeriform	150
" Gaseous	150
" how composed	150
" Liquid	150
" Solid	150
Brace	53, 116
Bracket	53, 116
Bright shafting	268
Brittleness	153

INDEX.

	PAGE		PAGE
Buoyant effect of water	219	Compressor, Air	235
Bushel, Cubic inches in	65	Concrete number	1
Butt joint	316	Cone	142
		" Altitude of	142
C.	PAGE	" Convex area of	142
Cancelation	21	" Frustum of	143
" Rule for	23	" Slant, Height of	142
Cantilever	261	" Vertex of	142
" Strength of	263	" Volume of	142
Capacity, Measures of	64	Consequent	97
Cause and effect	108	Conservation of energy	200
Center of gravity	160	Constants for cast-iron pillars	258
" " " of irregular plane figure	162	" " line shafting	270
		" " transverse strength	261
" " " " parallelogram	162	" " wooden pillars	259
" " " " regular plane figure	162	" " wrought-iron pillars	257
		Convex area of cone	142
" " " " solid	164	" " " cylinder	139
" " " " system of bodies	161	" " " cylindrical ring	146
		" " " frustum	144
" " " " triangle	162	" " " prism	139
Centrifugal force	194	" " " pyramid	142
" " Rule for	195	" " " sphere	145
Centripetal force	195	" " " surface of solid	139
Chains	251	Countershafts	268
" Rules for strength of	251	Couplet (ratio and proportion)	97, 100
" Treatment of	251	Cross-head friction in guides	193
Chamber, Air	244	Crowning of pulleys	171
Chord of circle	133	Crushing strength of materials	255
Cipher	2	" " " Table of	256
Circle	120, 133	Cube (solid)	138
" Arc of	121, 133	" of a number	77
" Area of	136	" root	79, 86
" Center of	120, 133	" " Proof of	92
" Chord of	133	" " Rule for	92
" Circumference of	120, 133	Cubic measure	63
" Diameter of	133	Cubical contents	139
" Divisions of	121	Curved line	119
" Radius of	133	Cylinder, Convex surface of	139
" Sector of	136	" " Volume "	139
" Segment of	137	Cylindrical ring, Convex area of	146
Circular ring, Area of	136	" " Volume of	147
Circumference of circle	133		
Coefficient of friction	189	**D.**	PAGE
" " " Table of	192	Decimal	38
Cold-rolled shafting	268	" number, how read	38
Combination of pulleys	184	" part of a foot, To reduce inches to	50
" " " Law of	185		
Common denominator	27	" point	38
" " Least	27	" To express as a fraction with a given denominator	52
Composite numbers	21		
Compound denominate numbers	62		
" lever	167	" To reduce a fraction to	50
" proportion	109	" " " to a fraction	51
Compressibility	152	Decimals, Addition of	40
Compressive strength of materials	255	" Division of	45

INDEX.

	PAGE
Decimals, Multiplication of	43
" Subtraction of	42
Degrees (circular measure)	121
" and limits of exhaustion	233
Denominate numbers	61
" " Addition of	68
" " Compound	62
" " Division of	73
" " Multiplication of	72
" " Reduction of	65
" " Simple	61
" " Subtraction of	70
Denominator	23
" Least common	27
Diagonal of parallelogram	129
Diameter of circle	133
Difference	9
" (in percentage)	56
Digits	2
Direct proportion	101
" ratio	97
Direction of force	156
Dividend	17
Divisibility	151
Division	17
" of decimals	45
" " denominate numbers	73
" " fractions	34
" Proof of	20
" Rule for	20
" Sign of	17
Divisor	17
Double belts	201
" shear	266
Double-acting pump	245
Downward pressure of water	209
Driver (pulleys)	173
Dry measure	64
Ductility	153
Duplex steam pump	246

E.	PAGE
Efficiency of machine	192
Elastic limit	248
Elasticity	152, 247
" Measure of	248
Energy	199
" Conservation of	200
" Kinetic	199
" Potential	200
Epicycloidal teeth	179
Equality, Sign of	4
Equilateral triangle	126
Evolution	79
Exhaustion of air	233

	PAGE
Expansibility	152
Exponent	77
Extension	151
" Measures of	62
Extremes of a proportion	101

F.	PAGE
Factor	21
" Prime	21
Figure, Plane	123
Figures	2
Fixed pulley	183
Follower	173
Foot-pound	197
Force	156
" Centrifugal	194
" Centripetal	195
" Direction of	156
" Point of application of	156
" Pump	242
" Representation of	160
" required to punch plates	267
Formula	115
" How applied	118
" Symbols used in	115
Fraction	23
" Improper	25
" Proper	25
" Terms of	25
" To invert a	35
" " reduce a decimal to a	51
" " " " whole or mixed number to a	26
" " " to a decimal	50
" " " " lowest terms	26
" Value of a	24
Fractions, Addition of	29
" Division of	34
" Multiplication of	32
" Reduction of	25
" Roots of	94
" Subtraction of	30
" To find least common denominator of	27
" " reduce to common denominator	28
Friction	189
" between cross-head and guides	193
" Coefficient of	189
" Laws of	190
" Table of coefficients	192
Frustum of cone or pyramid	143
" Altitude of	144
" Convex surface of	144
" Volume of	144

INDEX.

	PAGE
Fulcrum	165
Fundamental principle of machines	166

G.

	PAGE
Gain or loss per cent.	60
Gallon, Cubic inches in	65
Gallon of water, Weight of	65
Gas, Permanent	150
" Tension of	222, 228
Gaseous body	150
Gases, Application of Mariotte's law to	230
Gear-wheels	175
Gears, Bevel	175
" Horsepower of	204
" Miter	175
" Rules for	178, 180
" Spur	175
Gravity, Center of	160
" Specific	196
Guides, Friction on	193

H.

	PAGE
Hardness	152
Head (of water)	215
Helix	187
Hemp ropes	252
" " Strength of	252
Heptagon	130
Hero's fountain	237
Hexagon	130
Horizontal line	119
Horsepower	198
" of belts	202
" " gears	204
" " shafting, Formulas for	270
Hydraulic press	216
Hydrostatics	207
Hypotenuse	126

I.

	PAGE
Impenetrability	151
Improper fraction	25
Inclined plane	185
" " Rule for	186
Incompressibility of liquids	207
Indestructibility, Law of	152
Index of a root	79
Inertia	151, 158
Injector	239
" How operated	240
Inscribed angle	133
" polygon	134
Integer	2
Inverse proportion	101, 104
" ratio	97

	PAGE
Involute teeth	179
Involution	76
Isosceles triangle	126

K.

	PAGE
Kinetic energy	199

L.

	PAGE
Lacing belts	203
Lateral pressure of water	212
Law of Mariotte	229
" " Pascal	209
Laws of friction	190
" " inertia	157
" " liquid pressure	209
" " motion, Newton's	157
Lever	165
" Compound	167
" Fulcrum of	165
" Law of	165
Lifting pump	242
Like number	1
Line, Curved	119
" Horizontal	119
" Perpendicular to another	119
" shafting	268
" " Bearings for	269
" " Horsepower of	270
" " Rules for	271
" Straight	119
" Vertical	119
Linear measure	62
Lines, Parallel	119
Liquid body	150
" measure	64
" pressure, Laws of	209
Liquids, Incompressibility of	207
Local value of a figure	2
Long ton table	63
Loss per cent	60

M.

	PAGE
Machines, Pneumatic	231
" Principle of	166
" Simple	165
Magdeburg hemispheres	233
Magnitude of a force	157
Malleability	153
Mariotte's law	229
Materials, Crushing strength of	255
" Shearing " "	266
" Tensile " "	248
" Transverse " "	261
Matter	149
" Properties of	151
" Special properties of	152
Means (of a proportion)	101

INDEX.

	PAGE
Measure, Cubic	63
" Dry	64
" Linear	62
" Liquid	64
" of money	65
" " time	64
" Square	62
Measures, Miscellaneous	65
" of angles and arcs	64
" " capacity	64
" " extension	62
" " weight	63
Mechanics	149
Mensuration	119
Mercurial barometer	225
Minuend	9
Minus	9
Miscellaneous measures	65
Miter gears	175
Mixed number	25
Mobility	151
Molecule	149
Money, Measure of	65
" United States	65
Motion	153
" Newton's laws of	157
Movable pulley	183
Multiplicand	12
Multiplication	12
" of decimals	43
" " denominate numbers	72
" " fractions	32
" Proof of	16
" Rule for	16
" Sign of	12
" table	13
Multiplier	12

N.
	PAGE
Naught	2
Newton's laws of motion	157
Notation	1, 4
" Arabic	2
Number	1
" Abstract	1
" Composite	21
" Concrete	1
" Denominate	61
" Like	1
" Mixed	25
" Prime	21
" Reciprocal of a	97
" Unlike	1
" Unit of a	1
Numeration	1, 4
Numerator	23

O.
	PAGE
Obtuse angle	120

P.
	PAGE
Parallel lines	119
Parallelogram	123
" Altitude of	123
" Area of	124
" Center of Gravity of	162
" Diagonal of	129
Parallelopipedon	138
Parenthesis	53, 116
Partial vacuum	224
Pascal's law	209
Path of a body in motion	153
Pentagon	130
Per cent	55
" " Gain	60
" " Loss	60
" " Sign of	55
Percentage	55, 56
" Rules of	56
Perimeter of polygon	130
Permanent gas	150
Perpendicular lines	119
Pillars, Constants for	257, 258
" Formulas for strength of	256
Pinion	173
Pitch circle	177
" of gear teeth	177
" " screw thread	187
Plane figure	123
" " Center of gravity of	162
" Inclined	95
" " Law of	186
Plunger pump	243
Pneumatic machines	231
Pneumatics	222
Point of application of force	156
Polygon	130
" Area of	131
" Inscribed	134
" Perimeter of	130
" Regular	130
" Sum of interior angles of	131
Porosity	151
Potential energy	200
Power arm	165
" of a number	76
" " " ratio	99
Powers and roots in proportion	106
Press, Hydraulic	216
Pressure of air	223
" " atmosphere	223
" " liquid	209
" " water	215
" " " Downward	209

INDEX.

	PAGE
Pressure of water, Lateral	212
" " " Upward	211
Prime factor	21
" number	21
Principle of Archimedes	220
" " machines	166
Prism	138
" Convex surface of	139
" Volume of	139
Product	12
" Partial	15
Proof of addition	8
" " division	20
" " multiplication	16
" " subtraction	11
Proper fraction	25
Properties of matter	151
Proportion	100
" Compound	109
" Couplet of	100
" Direct	101
" Extremes of	101
" how read	100
" Inverse	101, 104
" Means of a	101
" Powers and roots in	106
" Rules for	101
" Simple	109
Pulley, Fixed	183
" Movable	183
Pulleys	170
" Balanced	171
" Combination of	184
" Crowning of	171
" Law of	185
" Rules for	172
Pump, Air	231
" " chamber for	244
" Force	242
" Lifting	242
" Plunger	243
" Steam	245
" Suction	241
Punching holes in plates, Force required for	267
Pyramid	142
" Altitude of	142
" Convex area of	142
" Frustum of	143
" Slant height of	142
" Vertex of	142
" Volume of	142

Q.

	PAGE
Quadrilateral	123
Quotient	17

R.

	PAGE
Radical sign	79
Radius of circle	133
Rate (in percentage)	56
Ratio	96
" Couplet of a	97
" Direct	97
" How expressed	96
" Inverse	97
" Reciprocal	97
" Symbol of	96
" Terms of a	97
" Value of a	97
" Velocity	189
Reading numbers	3
Reciprocal of a number	97
" " " ratio	97
Rectangle	123
Reduction of decimals to fractions	51
" " denominate numbers	65
" " fractions	25
" " " to decimals	50
Regular polygon	130
Remainder	9
Rhomboid	123
Rhombus	123
Right angle	120
Right-angled triangle	126
Ring, Circular	136
" " Area of	136
" Cylindrical, Area of	146
" " Volume of	147
Root	77
" Cube	79, 86
" Index of	79
" of gear tooth	177
" " ratio	100
" Square	79
Roots of fractions	94
" other than square and cube	95
Rope, Hemp	252
" " Strength of	252
" Wire	253
" " Strength of	253

S.

	PAGE
Scalene triangle	126
Screw	187
" Law of	188
" Pitch of	187
" thread	187
Sector, Area of	137
Segment, Area of	137
Semicircle	134
Semi-circumference	134
Shafting, Black	268
" Bright	268

INDEX.

	PAGE
Shafting, Cold-rolled	268
" Distance between bearings of	269
" Horsepower of	270
" Line	268
" Rules for	271
Shear, Double	266
" Single	266
Shearing strength	266
" " Rule for	267
" " Table of	266
Sign of addition	4
" " division	17
" " dollars	53
" " equality	4
" " multiplication	12
" " per cent.	55
" " subtraction	9
" Radical	79
Simple denominate number	61
" machines	165
" proportion	109
" value of figure	2
Single belts	201
Siphon	238
Slant height of cone or pyramid	142
Solid body	138
" Center of gravity of	164
Specific gravity	196
Sphere	145
" Area of surface of	145
" Volume	146
Spur gears	175
" " Horsepower of	204
Square	123
" foot	124
" inch	124
" measure	62
" of a number	77
" root	79
" " Rule for	84
" " Proof of	84
" " Short method for	85
Steam pump	245
Strain	247
Strength, Compressive	255
" of beams	262
" " belts	202
" " chains	251
" " materials	247
" " pillars, Formulas for	256
" " ropes (hemp and wire)	252
" Shearing	266
" Tensile	248
Stress	247
" Unit	247

	PAGE
Subtraction	9
" of decimals	42
" " denominate numbers	70
" " fractions	30
" Proof of	11
" Rule for	11
" Sign of	9
Subtrahend	9
Suction pump	241
Surface of solid	119
Symbol of ratio	96
Symbols of aggregation	53

T.

	PAGE
Table, Addition	5
" Multiplication	13
" of coefficient of friction	192
" " constants for cast-iron pillars	258
Table of constants for transverse strength	261
" " " " wooden pillars	259
" " " " wrought-iron pillars	257
" " crushing strength	256
" " distances between shaft bearings	269
" " shearing strength	266
" " tensile strength	249
Teeth of gears	177
" " " Addendum of	177
" " " Epicycloidal	179
" " " Involute	179
" " " Pitch of	177
" " " Root of	177
Tenacity	153
Tensile strength	248
" " of chains	251
" " " materials	249
" " " ropes	252
" " Rules for	250
" " Table of	249
Tension of gases	222, 228
Terms, Higher	25
" Lower	26
" of a fraction	25
" " " proportion	101
" " " ratio	97
Thread of screw	187
Time, measures of	64
Torricellian vacuum	234
Train	173
Transverse strength	261
" " of beams	261

INDEX.

	PAGE
Transverse strength of beams,	
Rules for	262
" " " cantilevers	263
" " " constants	261
Trapezoid	123
" Area of	124
Triangle	126
" Altitude of	129
" Area of	129
" Center of gravity of	162
" Equilateral	126
" Hypotenuse of	126
" Isosceles	126
" Right-angled	126
" Scalene	126
Troy weight	63

U.
	PAGE
Uniform velocity	154
Unit	1
" of a number	1
" square	124
United States money	65
Unlike number	1
Upward pressure of water	211

V.
	PAGE
Vacuum	224
" Partial	224
" Torricellian	234
Value of a fraction	24
" " ratio	97
Vapor	150
Variable velocity	154
Velocity	154
" ratio	189
" Uniform	154
" variable	154

	PAGE
Vertex of angle	120
" " cone	144
Vertical line	119
Vinculum	53, 116
Volume of circular ring	147
" " cone	142
" " cylinder	139
" " cylindrical ring	147
" " frustum of cone	144
" " " " pyramid	142
" " prism	139
" " pyramid	142
" " sphere	146
" Unit of	139

W.
	PAGE
Water, Buoyant effect of	219
" Gallons in a cubic foot of	65
" Incompressibility of	207
" Pressure of	209
Wedge	187
Weight	151
" Arm	165
" Avoirdupois	63
" lifter	234
" Measures of	63
" Troy	63
Wheel and axle	169
Wheelwork	173
Wheelwork, Rules for	174
Wire rope, Strength of	253
Work	197
" Measure of	197
" Unit of	197
Worm and worm-wheel	176

Z.
	PAGE
Zero	2

www.ingramcontent.com/pod-product-compliance
Lightning Source LLC
Chambersburg PA
CBHW031857220426
43663CB00006B/664